49 ⁵⁰

COULTER LIBRARY ONONDAGA COMM. COLL.

3 0418 00219819 8

QL
82   Benirschke, Kurt
.B46
1986   Vanishing animals

DATE DUE

D1608545

QL   BENIRSCHKE, KURT   09/17/87
82   VANISHING ANIMALS
.B46 (2) 1986 .
1986 0165 03 174966 01 4   (IC=0)

SIDNEY B. COULTER LIBRARY
Onondaga Community College
Syracuse, New York 13215

Vanishing Animals

Vanishing Animals

Art

Andy Warhol

Text

Kurt Benirschke

Springer-Verlag
New York Berlin Heidelberg London Paris Tokyo

Library of Congress Cataloging in Publication Data
Benirschke, Kurt.
  Vanishing animals.
  Bibliography: p.
  I. Rare animals.   I. Warhol, Andy.
II. Title.
QL82.B46   1986   333.95'416   86-21983

Many photographs included in this book were taken by Ron Garrison, Staff Photographer, San Diego Zoo.

Text © 1986 by Kurt Benirschke.
Cover art, color art, and art on pages 3 and 7 © 1986 by Andy Warhol.
Photo of Andy Warhol on page vii © Richard L. Schulman.

All rights reserved. No part of this book may be translated or reproduced in any form without written permission from the copyright holder.

Typesetting by David E. Seham Associates Inc., Metuchen, New Jersey.
Printed and bound by H. Stürtz AG, Würzburg, Federal Republic of Germany.
Printed in the Federal Republic of Germany.

9 8 7 6 5 4 3 2 1

ISBN 0-387-96410-X Springer-Verlag New York Berlin Heidelberg
ISBN 3-540-96410-X Springer-Verlag Berlin Heidelberg New York

Contents

Introduction 1
California Condor 10
La Plata River Dolphin 16
Mouse Armadillo 22
Mongolian Wild Horse 28
Giant Chaco Peccary 34
Okapi 40
Whooping Crane 46
Komodo Monitor 52
Sumatra's Rhinoceros 58
Douc Langur 64
Bats 70
Butterflies 76
Sömmerring's Gazelle 82
Galapagos Tortoise 88
Puerto Rican Parrot 94
Further Reading 99

Andy Warhol

Kurt Benirschke

Extinction—the tragic and permanent loss of entire species of animals—should be a concern for everyone. This concern and a strong desire to take action toward preventing the loss of more animals has brought about an unusual collaboration between art and science. The result is this beautiful volume in which artist and scientist have joined efforts to inform and inspire others to take action.

Especially for this book, Andy Warhol has created prints (silkscreen over collage) of some of the most endangered animals in the world. Here they are joined with a stimulating text by Dr. Kurt Benirschke affording the reader an opportunity to discover the lives and habits of these animals—and what the outlook is for their survival.

Andy Warhol is one of today's most creative artists, widely known for his highly innovative portraits and representations of popular culture. In 1983, Warhol expressed his love of animals and concern for the environment in a series of 10 prints of *Endangered Species*. For this volume, he has returned to this subject, creating 16 new prints.

Kurt Benirschke, M.D., is well known for his work in pathology and widely published in that field.
Benirschke's interest in conservation is more than a "hobby," one to which he is extremely devoted. For many years he has helped to rescue endangered species through breeding in captivity and conservation in the wild. Benirschke is Professor of Pathology and Reproductive Medicine at the University of California, San Diego, and is Trustee and former Director of Research of the San Diego Zoo.

It is hoped that these fascinating and striking portrayals will stimulate readers to join their own energies and talents to this important fight against the loss of more species.

Douc Langur

Introduction

What is man without the beasts? If all the beasts were gone, men would die from a great loneliness of spirit. For whatever happens to the beasts, soon happens to man. All things are connected.

So spoke Seattle, Chief of the Dwamish and Allied Tribes of Puget Sound, in 1855, in reply to "the Great Chief in Washington (who) sent word that he wishes to buy our land. How can you buy or sell the sky, the warmth of the land? The idea is strange to us," he said.

These were prophetic words a century ago. Nature was still very much at equilibrium. To be sure, the Dodo from Mauritius, a flightless bird, had been exterminated by man 200 years earlier, the great auk of the Atlantic seashores had just lost its last representative, and the billions of North American passenger pigeons were being reduced by hundreds of thousands annually (Martha, its last member, died in 1914). But, by and large the threat to animals was not yet comprehended, except, perhaps, by such Indians as Chief Seattle, who had seen nearly all the buffalo shot by early settlers. And by men like John Muir, who was responsible for the establishment of Yosemite Park and other sanctuaries. These men foresaw the great threat of the future—an expanding human population having ever greater needs and desires. Better equipped as we are, we have become the threat to this planet's wildlife, plants, and water, and the air we breathe.

Since the explorers of the last century brought the knowledge of the great wonders of the world and the beauty of unseen animals and planets into our view, much has changed. So much, in fact, that Pulitzer Prize winner Edward O. Wilson exclaimed in 1980 that "The one process ongoing in the 1980's that will take millions of years to correct is the loss of genetic and species diversity by the destruction of natural habitats. This is the folly our descendants are least likely to forgive us."

Extinction has been an ongoing process for millenia. However, it was a natural process in the past. As animals and plants adapted to changing environments, their ancestors vanished, but succeeding forms evolved and life went on, and it has gone on for thousands of plants and animals for millions of years. Not so any longer. The process of extinction has increased dramatically, in parallel with the expansion of the human population, and if the evolution of animals is proceeding—it cannot be watched—it can no longer catch up with extinction. Now there is danger that this is the end of the line for many beasts, and with this realization comes our reaction, the wish to do something about it. We examine the problem, we spread the word, and then we try to change the course of events to come.

This book intends to bring some of the less well known endangered animals to the reader's attention. These animals deserve just as much attention as the giant panda or the mountain gorilla about which so much has already been said.

Such animals as the panda and the gorilla may be regarded as "flagship species" that draw attention to the underlying problem: the wanton destruction of habitat, of food plants, and of suitable refuge. But also they serve as examples of how international concerns wrestle with the problem and how last-ditch efforts endeavor to stem the tide. Naturally, the animals presented here are very personal choices, having been selected from a virtually endless supply of animals whose last hour is rapidly approaching.

The International Union of Conservation of Nature (IUCN) with headquarters in Switzerland publishes its *Red Data Book* of mammals, amphibia, plants, invertebrates, and other taxa. Through its large network of specialists, the IUCN collects in these books information on all life forms and alerts the public to the degree of endangerment. They assess the causes of the threat to their survival and make recommendations to ensure safeguards. At a meeting of these scientists in 1985 they listed the 12 most endangered animals—as well as plants. Strict criteria were followed in their selection, such as urgency, prominence, and biological value of the many species under consideration. But they also faced an "embarras de richesses," that is, there were too many species to pick from. But they chose the following animals as the most endangered:

1. Bumblebee bat—discovered only in 1974 and the smallest mammal.
2. Kouprey—only few if any of this Indochinese cattle-like animal live.
3. Mediterranean monk seal—fewer than 500 survive and it is already extinct in most of its former habitat.
4. Woolly spider monkey—America's largest monkey, deprived of its habitat by expanding Brazilian cities.
5. Pygmy hog—only few survive in the Himalayan foothills.
6. Northern white rhinoceros—perhaps 10 still live in Zaire.
7. Sumatran rhinoceros—this book tells of its last survivors.
8. Madagascar tortoise—none exist in the wild; the remaining few are in private hands.
9. Orinoco crocodile—hide collection has brought it to near-extinction.
10. Kagu—a flightless bird from New Caledonia.
11. Hawaiian tree snails—more than half of the 41 species have become extinct already.
12. Queen Alexandra's birdwing butterfly—the world's largest butterfly from Papua.

Sömmerring's Gazelle

There were many runners-up and, like the 12 chosen species, they often were not very "glamorous" species such as the panda. Such animals are the innocent bystanders in a deforestation process that claims 27,000,000 acres of tropical forest annually. They form the web of life of which we are but a part. "Extinction is forever—all things are connected!" Indeed, much can be said for the idea that many of these less glamorous species perform a disproportionally larger task in this web of life. Think of the pollination of plants, much of which is accomplished by insects such as bumblebees, butterflies, and the like. The notion that the abundance of butterflies is an excellent indicator of the state of the environment is not too far-fetched. The disappearance of butterflies as a result of use of pesticides and the loss of its foodplants because of prevalent monocultures has not gone unnoticed. CITES, the treaty signed by many countries that seeks to stem the trade in endangered species and its parts, has long prohibited the trading of precious butterflies for collectors. Yet as a result of these restrictions, the ingenuity of man has sought to overcome them. In Papua, New Guinea, for instance, home of the endangered Queen Alexandra's birdwing butterfly, the farming of butterflies has become a profitable business. It provides employment, has led to reafforestation with native foodplants, and an increase in the butterfly populations. Some specimens can now be selectively and legally exported. Similar events have taken place in Europe and other parts of the world. Once a problem is recognized to have been created by man's overabundance—there are now 5 billion of us, a population tripled in the last century—we tend to seek its correction, and often at a considerable price.

The zoological community has been similarly affected by the disappearance of species. Barely a hundred years old, zoos have needed to change their entire philosophy in recent years. With the creation of the Endangered Species Act in 1973 it was no longer possible to stock the cages as before. Trading became ever more difficult, trappers in foreign lands disappeared, and many animals vanished at the hands of ivory collectors and hide traders. They were also displaced by an expanding population with its cattle herds and agricultural needs. Zoos had to reexamine their priorities if they wished to continue existence. A new spirit of cooperation developed, studbooks of rare animals were created, breeding loans effected, and, for the first time, the zoological community became conscious that their animals represented treasured charges that needed to be propagated as though no new animals would ever come from the wild. This meant difficult choices—should we breed Siberian or Sumatran tigers, African or Asian elephants, since clearly not even all zoos together can look after all of the 4,000 species of mammals, let alone the even more numerous birds and reptiles. Through many meetings, choices were made, the species survival programs established, and strategies were mapped. It was soon realized, however, that not enough was known about the biology of the animals they had cared for to create self-sustaining populations, and an intensive research effort was launched. Genetics, reproduction, diseases, nutrition, and many other topics were explored to enhance survival and perpetuation of animals in captivity. The chapters in this book will present some of the interesting problems that have been encountered in this massive effort.

Why should there be zoos at all? The question has been asked more often and more urgently in recent years. Surely zoos cannot rescue all species from extinction. They have been able to do so for Przewalski's horse, for the Arabian oryx, and they are making a concerted effort to save the California condor from oblivion. As a result of successful zoo management there have been so many Arabian oryx bred that reintroduction into their original habitat of Oman became possible in 1982 and already second-generation offspring have been born. Still, only a few species can be saved through the efforts of zoos and, therefore, the question of "ex situ" (captive) versus "in situ" (natural habitat) conservation has become the focal point of our time. Realizing the relentless downhill trend of natural habitats, zoos have formed strong alliances to do their best with ex situ conservation. Above all, they have endeavored to educate the general public and to learn about the basic biologic properties of animals. The "in situ" conservationists, of course, have a much stronger suit. Who can argue with the fact that parks, reserves, and indeed original habitats, are not only much less expensive ways to conserve, but also conserve many more different taxa of plants and animals at the same time. On the other hand, most of these habitats are inaccessible in foreign lands, and it would require enormous conservation dollars to safeguard them. Moreover, they are under constant pressure by the expanding populace. Our conviction is, then, that we must have both aspects of conservation activity, ex situ as well as the in situ conservation of parks and reserves. But how can all of this best be achieved?

On a global level the major conservation agencies attempted to bridge the gaps between individual efforts when they launched the World Conservation Strategy in 1980. This group seeks to combine conservation with development, stating that "the underlying problems must be overcome if either is to be successful." To be sure, these are highly controversial topics that have been addressed in numerous conferences and books. And there is no universally acceptable solution. But we must continue to try, because all agree that life on earth will not only be much less pleasant, but certainly much more expensive when plants and animals are destroyed. We must not delude the public that artificial management (of insect pests by pesticides when the bats are gone) will be possible or easier, or that food resources can be readily multiplied to sustain additional billions of people. It will be expensive if at all possible. I often wonder when I see *Star Wars*, and similar movie spectaculars, about how our children envisage that these warriors grow their plants or food animals to sustain their lives. Frequently, we extol that cryopreservation, the freezing away of seeds, embryos, and the like, will save existing genetic diversity from oblivion. How unrealistic are such hopes when we don't yet have a clue of how to get the semen, let alone the fertilized embryo, for safekeeping from such forms as the blue whale, the bumblebee bat, or the many other animals and plants that are presently threatened. "Sustaining Tomorrow," a recent review of the Strategy for World Conservation and Development, gives an excellent insight into the complexity of our current problem. No doubt education and research are the commands of the day, with a leveling of population growth presently being the only hope we can envisage for a sane future. And we are in this together; no one escapes.

As far as animals are concerned—and, after all that is the topic of this book—we need to be mindful of the possibility that "Tomorrow's Ark is by Invitation Only." Will we have the wisdom to look after the hyena and the bat, the way we concern ourselves with the severely threatened mountain gorilla of Rwanda or China's giant panda? And who will make the choices of what will be conserved?

Very often such choices are made through the efforts of individuals. Take the bats, for instance. So rapid was the decline of bats, so vicious was Western man's pursuit of what he perceived to be a harmful animal, that Dr. Stebbings in England and Dr. Tuttle in the USA formed the group "Bat Conservation International" in 1982. After all, of some 4,000 species of mammals, bats, with 942 species, are a huge and diverse group, second in number only to the rodents with 1,300 members. This conservation group has most successfully begun to educate the public on the benefits we derive from bats (insect predation, pollination, seed dispersal) and has been instrumental in preserving bat caves and constructing bat roosts.

The late Dian Fossey long prevented trophy hunting of mountain gorillas; George Schaller sought the causes of threats to the giant panda, and others engage their efforts, indeed subvert their entire lives, to the rescue of the manatee, the black-footed ferret, or the whooping crane. Individual scientists have taken on the challenge to

Mongolian Wild Horse

preserve peregrine falcons, the Chaco peccary, the California condor, and so on. There is no shortage of individuals to attempt rescue operations but, by and large, such efforts come in the later years of the species, at a time when they have already come close to extinction. On a global scale it is the IUCN with its specialists' groups that seeks an overall coordination of efforts. The reader will perhaps be interested to obtain a copy of the periodic Newsletter of the Species Survival Commission from the IUCN at Gland, Switzerland, and he or she also would benefit from reading Oryx, the journal of the Fauna and Flora Preservation Society. These publications, more than any others, will provide an insight into the problems facing various animals and efforts to overcome them.

The interaction of zoos in this global crisis at first might be considered to be only miniscule. After all, their "carrying capacity" is finite and not all animals can be managed well (let alone profitably) in zoos. Soulé and others have recently addressed this question in "The Millenium Ark: How Long a Voyage, How Many Staterooms, How Many Passengers?" (Zoo Biology 5:101, 1986). They conclude that some 2,000 species of vertebrates (other than fish) will need this protection in the future—a seemingly overwhelming task, and the first order of need is an increase in our knowledge of these animals so that they can be managed successfully. And zoos will have to jointly agree on who saves what species and stick by their resolutions. Fortunately, these agreements are being forged with increasing rapidity and sophistication. Computerization and biologic research also proceeds apace to make the programs successful.

What is needed most urgently is a greater commitment to the task. The former President of the Zoological Society of San Diego, Sheldon Campbell, examined the requirements in a paper entitled "Before We Pilot the Ark." He says, "what people care for most they spend money on, for in our society, value is easily measured by the dollars we spend: we value freedom and spend billions for atom bombs and rockets; we value emotional highs and spend billions on alcohol and drugs; we value personal security and spend billions on law and order; we value entertainment and spend billions for rock music concerts, baseball games, and video tapes; we value health and spend billions for cancer research and vitamin pills." We must be future-conscious, he concludes, and we must become educated to the needs of conservation. That is probably the overriding beneficial effect that proper zoo exhibits, and all that goes with them, have on society—they make us sophisticated enough so that the value of wildlife becomes as valuable to us as, for instance, rock concerts.

We must create an environmental ethic for the benefit of our children. Just how complex such an issue is can be demonstrated by observing the plight of America's largest bird, the California condor. When the decision was made to take the remaining few into captivity so that their relentless downhill course might be reversed, one group of conservationists challenged the action in court. "May they die in honor, for a captive bird represents only its genes and feathers, they are not possessed with majesty of the soaring animal we have come to revere," say these well-intentioned individuals. The counterarguments are explored in our chapter on this vanishing species, and the courts have seen the wisdom in this instance, recently affirming that ex situ conservation seems to be a practical goal.

There are no easy answers to the many questions that now face us so suddenly. We are ill-prepared to make so many decisions so suddenly—yet there is no other way. We must make educated choices now. This book will provide a small insight into the issues that are faced by conservationists currently working in zoos. It hopes to bring some of the less well known animals closer to your heart. Above all, it is written in the spirit that

We have not inherited the Earth from our parents, we have borrowed it from our children.

In considering the plight of the California condor—only 26 are alive at the time of this writing—we face the most important issue of contemporary conservation: Is "ex situ" conservation useful, or is it true that animals that live only in zoos do not "live?" Well-intentioned and strongly determined though both sides in this controversy are, we believe it to be a travesty of our intentions to try and settle the dispute in the courts, as though they possess the wisdom needed in this critical situation. In fact, if court intervention had not prevented the capture of a condor pair in 1950 by Belle Benchley, then director of San Diego's Zoo, it is possible that it wouldn't be necessary to write about their plight now. For she was able to prove that a related species, the Andean condor, was breeding four times more efficiently in the zoo than in the wild. Why not give the California condor the same chance?

California Condor

(*Gymnogyps californianus*)

Gymnogyps californianus takes its name from the Greek *gymnos* for naked (its head) and *gyps* for vulture. It is our largest bird, with a wing span of up to 9½ feet and weighing from 20 to 31 pounds. Fossil records from Florida and the Los Angeles La Brea tar pits indicate that an essentially similar bird existed during the ice ages some 40,000 years ago, and that its range at that time extended from the Columbia River to Baja California and east through the southern United States to the Atlantic Ocean. When Lewis and Clark came to the Columbia River in 1805 they described many "beautiful buzzards of the Columbia" which fed on washed up salmon. In sketching these birds they accurately depicted the head of the condor. However it was first recorded by a Carmelitan friar in 1602 who saw a flock feeding on the carcass of a whale. And from all the regions it once inhabited it has now retreated to a V-shaped area in the southern mountain ranges of the San Joaquin Valley. It has not been seen in Monterey since 1861 and was last seen in Baja in 1937.

The exact causes for the condor's rapid decline remain a matter of dispute. It is inedible and was hunted only for the purposes of private and museum exhibition. It has often been said that early settlers used its large quills to transport gold dust, but this story remains unsubstantiated. Eggs have been removed from nests for study and exhibit, but while this may have had some impact on such a slow breeding species it hardly suffices as the cause for the rapid and drastic decline in the condor population. Also revered by the area's Indians as a symbol of immortality, it was sometimes captured and used in various tribal rituals, but once again the numbers of birds used for such purposes were small and hardly account for the present dilemma.

Ultimately the causes of the condor's decline doubtless came from modern man's incursion into its habitat with guns and poisons. Many were shot simply for "sport" because they, like the 21 other existing species of vultures, were declared vermin. Then there was the rapid decline of its customary food sources, elk and pronghorn, these being decimated by the rifles of an expanding human population. Some condors broke their wings in high-tension wires. Others ate carrion containing poison. Although documented only once in 1974, an immature bird was found to have high levels of DDT/DDE in its blood. This insecticide is known to accumulate toward the end of the food chain and is also known to disrupt reproductive processes. Indeed the last egg obtained from a condor nest in the spring of 1986, taken from the last wild

(Above) Large flight cage, the "condorminium," for the San Diego Zoo's San Pasqual breeding program. One youngster is seen in the middle of the picture. Eight visiting turkey vultures are frequent guests and look at their relatives from above. (Ron Garrison, San Diego Zoo)

female, Adult Condor-8 (AC-8), was abnormal. Its shell was 58 percent thinner than normal, a well-known effect of insecticide intoxication. Because of its thinness it broke, perhaps under the weight of its mother, and spoiled. Other birds have died from lead poisoning. One of the last females to be captured (AC-3) died in January 1986 despite strenuous efforts to remove the lead from her blood. She was so weak when captured that she offered no resistance and was unable to digest food, as her

(Above) First captive-hatched (March 30, 1983) California condor, a male named Sisquoc, being hand-fed by specially made puppet to avoid imprinting on man. (Ron Garrison, San Diego Zoo)

muscles were so weakened by the lead. While she had eight lead pellets in her muscles as evidence of having been shot at, it was the pellet contained in her stomach that caused her death. This pellet must have come from feeding on a deer carcass that had been killed by a hunter's shotgun. Only when the stomach acids dissolve the lead can it enter the animal's bloodstream and become harmful or lethal. Another wild bird was found to have ingested cyanide used for killing coyotes. No animal, however well adapted for survival, can withstand these pressures.

And on top of all this, the condor has, as we have already mentioned, a notoriously slow reproductive process. It does not reach sexual maturity until it is seven years old, reaches a maximum age of 40, and lays an egg only every other year. Under the very best conditions then, a female may have 16 offspring. But squabbles between the parents, predators (notably ravens), and occasionally even inclement weather may raise havoc with incubation or destroy the egg. Thus, juvenile mortality is high. Finally, California condors mate for life. They have strong pair-bonding qualities, and when a mate dies it actually spells the reproductive end of two animals. All these factors combined have had serious impact on the condor population. In the past quarter century their numbers have fallen 10-fold to the point where only four animals remain on the wing. And all this has taken place despite the fact that the condor was declared a protected species as early as 1905.

The flight of the condor is perhaps its most exhilarating aspect. Once airborne—and this may take some doing from flat ground—it can soar seemingly without effort for hours on thermals and windcurrents, rarely needing to flap its wings. It may reach speeds of 80 miles per hour and often covers hundreds of miles a day in search of carrion.

Condor young hatch from a four-inch greenish-white egg that is laid on bare ground in totally inaccessible caves or on rocky mountain ledges. The 42–60-day incubation is undertaken by both parents. The hatchling has a whitish gray down, weighs about 7 ounces, and fledges after some 6 months of parental care. It retains its pale color for some 2 years and depends heavily on regurgitated food from its parents for many months. We have learned to sex the birds by chromosomes that are prepared from tiny blood samples taken from their wings.

In order to prevent the "imprinting" on bird keepers (as is so well known from Konrad Lorenz's geese), we treat captive hatchlings quite specially. They are raised in isolettes that were originally designed for premature babies. The hatchling is fed chopped meat by a hand that is clad in a condor head puppet. Later they receive regurgitated food from turkey and king vultures so as to simulate nature as much as possible. They are nearly insatiable eaters and gain as much as 10 percent of their body weight each day!

It has been found that, as with other birds, "double-clutching" is possible. Normally the condor lays only one egg in March or April, but when this is destroyed during a squabble or by a predator, the female will lay again. By means of this methodology it has been possible to induce the same female to lay year after year, without a period of rest. Thus egg production can be markedly increased. And that is the hope and intent of the captive breeding program: to produce as many birds as quickly as possible so that none of the currently available genetic background will disappear. These efforts began in 1975 with the "Recovery Plan" that was first enunciated by Sanford Wilbur. Two large sanctuaries (Sespe 1947, Sisquoc 1973) had been set aside following the advice of the earliest student of the species, Charles Koford, who had estimated that in 1953 only 60 animals were left, and who had also predicted the rapid decline in their population. The Endangered Species Act of 1973 focused renewed interest on the condor. Solar-powered transmitters were attached to designated pairs, and their flight pattern, nesting areas, feeding frequency, and other behavioral habits were monitored. A telemetry-outfitted swan's egg was exchanged for a condor egg to ascertain the frequency of egg turning and incubation temperature, and all condors were extensively photographed thus making it possible to specifically identify each bird, to discern its habits and lineage, all of which is invaluable information for the construction of a sound management program.

Meanwhile it was found, through observing the similar (though less endangered) Andean condor, that "released birds" could be entrained onto relatively small areas of territory as long as food was available. Michel Terrasse in France had found the same to be true for the European griffon vulture. He had released birds into areas where they had become extinct. He found that contrary to belief they did not need "guide birds" for survival in the wild. They had low mortality, established a territory of 100 km^2, and nested near the aviaries from which they had come.

Currently, through the untiring efforts of Noel Snyder, 13 California condor chicks have been hatched during the last 3 years. With the additional eight that have been trapped during the same time, the total population is now four in the wild and 23 in captivity. This represents a decided upward trend. With some luck, breeding will commence by 1990 and if double-clutching succeeds an even larger population can be expected soon. And while all of the genetic diversity currently available is already represented at the San Diego and Los Angeles Zoos, future management may require the use of artificial insemination in order to mix these genes.

So, as Robert Redford said in his recently released condor film, "Don't call us fools!" And don't imagine that all these activities are merely "a zoo's gimmick" to acquire some new birds for exhibition. Call our efforts what they truly are—a last-ditch effort of conservation of North America's largest bird, the symbol of immortality.

Whales, porpoises, and dolphins all belong to the order Cetacea (Latin for large sea creature). There are some 80–100 members of this order, the number being dependent on whether the taxonomist is a "lumper" or a "splitter." Among the Cetacea there are two main subgroups, the toothed whales (Odontoceti), and the baleen whales (Mysticeti), and a fierce battle has raged among scientists in their efforts to decide whether these two groups came from one land mammal ancestor or whether each had its own progenitor. Resolving the question is particularly difficult as whale fossils are relatively sparse, but for the moment current opinion favors a monophyletic origin with separate lineages developing *after* these mammals took to the water. The word porpoise is derived from the Latin *porcus* (pig) and *piscis* (fish). The Romans considered it a

La Plata River Dolphin

(*Pontoporia blainvillei*)

"seapig" and cherished its succulent meat as food. The Greeks, on the other hand, revered dolphins. Aristotle compared them to man himself. The word dolphin is derived from the Greek *delphys*, meaning womb, and this designation follows them even today as they comprise the Cetacean family Delphinidae. The difference between dolphins and porpoises (considered beakless) is now blurred and it is best not to make sharp distinctions. They both belong to the order Cetacea, and their various characteristics allow us to place them into various suborders and families. The best-known feature of whales and dolphins is, of course, their intelligence. This can be inferred not only from the ease with which they are "trained" and from their social behavior, but also from extensive studies of their brains, which are unusually large and have a highly complex folded surface that is not found in most other mammals. A question that has often been debated, but that has yet to be resolved, is whether or not whales and dolphins sleep. We believe that sleep is necessary for all animals, but is protracted sleep as we know it necessary or is the dolphin's occasional "catnapping" sufficient? These mammals have to surface frequently for air, and rough seas may not allow slumber without inhalation of water through the blowhole (which is not connected to the mouth as it is in other mammals). Some authorities have suggested that one half of the brain may sleep while the other keeps the animal going. Yet in a study of a Ganges River dolphin conducted by Giorgi Pilleri, a noted Swiss researcher, sleep apparently did not occur, for the animal studied swam continuously around the clock. So this seemingly simple question remains unanswered.

River dolphins or Platanistidae (a word referring to their flattened beaks) belong to the toothed whales and are remarkable in many ways. Unlike other odontocetes for instance, the La Plata River dolphin does not have the characteristic asymmetrical skull of other toothed whales, which comes about through the atrophy of one nasal cavity. Also, in contrast to the 100 or so teeth of most toothed whales and dolphins, river dolphins have 200–250 fine, very pointed teeth and their lower jaw, which is more mobile than the jaws of other toothed whales, is fused in midline for a long distance. They dive for only a few minutes at a time, their eyesight is relatively reduced or absent, they possess peculiar plates of bone in their skin, and have many other characteristics that suggest similarities to the oldest whale fossils on record. They are, therefore, considered to be the most "primitive" examples of whale-like animals. But it seems as though perhaps their time has come, for while fossil records indicate that river dolphins at one time existed in many parts of the world, today these "sweet-water" adapted Cetacea exist only in South America, India, and Asia.

Lipotes vexillifer, the Chinese or Yangtze River dolphin, is about 7–8 feet long, weighs around 90 kg, and has a gray back and white belly. At one time also known as the Tung Ting Lake dolphin (as it was found in that lake), it is now extinct in that area and has suffered marked reduction in the river due to fishing, motorboat accidents, dam building, and other human activities. In Pakistan's Indus River and its tributaries lives the essentially blind Susu (*Platanista indi*), while the also virtually sightless *Platanista gangetica*, a very similar species, comes from the Ganges River. They are both severely threatened by fishing activity and by dam building which, during irrigation of adjacent fields, also isolates populations of dolphins making them even more susceptible to fishing. The creation in 1979 of a sanctuary for whales in the Indian Ocean which includes the Indus and Ganges Rivers has probably been an essential effort aimed at conservation of these animals.

The Inia (Bolivian for dolphin) is the best known of the river dolphins. It lives in the great Amazon and Orinoco River systems as Inia geoffrensis and in the Bolivian Madeira River as Inia boliviensis. These two dolphins are very similar and have often been exhibited in delphinaria. All river dolphins (and particularly Inia) have mobile (as opposed to rigid or fused) necks and upper chest vertebrae as do all other odontoceti. Both dolphins are light pink, measure up to 3 meters, and can attain weights of over 100 kg. In contrast to many other river dolphins, Amazon River dolphins can see rather well. They are said to aid injured conspecifics by lifting them to the surface and are also said to follow boats that contain their captured young.

The La Plata River dolphin derives its name, Pontoporia blainvillei, from the Greek word for "seaferrying" and from the name of the French professor of zoology, de Blainville. Known as the "tonina" in Brazil, it is called the "Franciscana" in Uruguay because the distinct brown color of its young looks like the brown cloaks worn by the local Franciscan monks. As it matures, it becomes light gray. La Plata River dolphins are the only river dolphins that also go to sea. They are found in the brackish water of the mouth of the La Plata River and often as far as 10 miles out at sea. Their main peril comes from shark-fishing activity that is based in several small Uruguayan ports. These fisheries produce "bacalao nacional," an item previously imported from Norway. Bacalao is essentially shark meat cut into long slivers that are then salted and dried. During Lent it is a welcome substitute for the large quantities of beef that would otherwise be eaten. Uruguay's coastal waters are unusually rich in several species of large shark and the many species of fish and other marine fauna upon which the sharks feed. The fishermen plant their large nets (made of 2 mm nylon threads with squares of 20–30 cm) near the ocean floor and leave them overnight or longer. It appears that Pontoporia feed from

(Below) Skull of river dolphin, adult male Pontoporia blainvillei. Over 200 needle-shaped teeth can easily be counted. Note that the lower jaw is fused over most of its length. The blowhole can be seen on top.

the same schools of fish and travel in harmony with the sharks and so they are accidentally caught in these "trammer" or "gill" nets, and they drown. Initially no use was made of the dolphins captured in this way and they were simply discarded. More recently though, oil has been made by boiling their blubber. This oil is used for waterproofing and motoroil. The meat may be fed to pigs. It has been estimated that between 1,000 and 1,500 La Plata dolphins die annually as a result of these fishing activities. But whether this has any significant impact on their total population is unknown, since it has been impossible to estimate the size of the population in brackish water.

It has often been remarked that it is odd that *Pontoporia* should be caught in such nets inasmuch as this species of river dolphin possesses the largest eyes and optic nerves of its group and should therefore have relatively good vision. Also, river dolphins have an efficient echolocating system that is designed to catch fish and small crustaceans. Some dig up such creatures in the mud of river floors with their long beaks, presumably after echolocating them. Therefore, one would assume, they should be able to detect the nets and shy away from them. But perhaps, so it is thought, they are traveling too quickly and detect the net too late to avoid it. When nets of 10-cm squares are used, no dolphins are caught, presumably because they are detected. But since these nets are less efficient for shark fishing, they are not used. Gill net fishing has, however, allowed another challenging observation. Giorgi Pilleri detected small skin defects in a young animal that had drowned in a gill net. He interprets these defects as resulting from attempts by other dolphins to rescue their young. Whether this interpretation is correct is open to debate, but it certainly is in keeping with the nurturant behavior displayed by the Amazon dolphins as described above. Be that as it may, many dolphins perish needlessly. There is some hope that things may improve as recent prosperity has enabled some fishermen to purchase stronger, more seaworthy boats, and now shark fishing takes place farther out at sea where the dolphins travel less commonly. Thus the annual dolphin kill has been declining, according to R. Praderi, Uruguay's best-known cetologist.

Yet the 1986 whaling ban does not apply to these smaller cousins of the whales, and some authorities fear that this may have a devastating effect on these animals inasmuch as their meat and oil may be used as a substitute in many cultures when whale meat is no longer available.

In the sixteenth century, when Spaniards exploring the New World first encountered this curious mammal, they named it "armadillo" or, loosely translated, "little tank." Armadillos are grouped with ant-eaters in the order Edentata, meaning "toothless," and for the most part they are just that, except for a row of tiny enamel-free teeth that they use to hold onto the grubs and insect larvae they eat. Yet because they are nocturnal creatures, we still know precious little about them.

Mouse Armadillo

(Chlamyphorus retusus)

One of the finest natural history museums in South America is located in the magnificent town of La Plata, 30 km south of Buenos Aires. Here statues of saber-toothed tigers keep a watchful eye over the entrance to halls filled with the bones of enormous prehistoric animals, dug from their graves in the La Plata River delta. Here also is a fine exhibit of the various armadillo species still living in South America. One can clearly see their monstrous ancestry in the shapes of their peculiar shell components or scutes. In fact most armadillos are named mainly for their extraordinary external markings, as we know so little else of their secret nocturnal lives. There is the nearly extinct giant armadillo, the naked-tailed six-, seven-, three-, and nine-banded armadillo and the bolita, a three-banded species of the Chaco. Last but not least, there are the two most adorable of the species, the "mouse" or "hairy" armadillos, so named for their furry undersides.

A remarkable but sad story will illustrate how profoundly ignorant we still are, after almost four centuries, where these relics of a by-gone era are concerned. Some years ago the curator of one of Europe's great zoos was kind enough to permit me to take a tiny skin sample from the zoo's giant armadillo. At the time, my interest in chromosomal speciation demanded the establishment of a tissue culture from such an animal. The zoo was the only one in the world with such a captive jewel. It was not easy to persuade four reluctant zookeepers to assist me in turning this giant creature (60 kg) onto its back

(Left) Armadillo precursor in the Museum of La Plata.

while it slashed at us with its powerful claws that are designed to break open cement-like termite hills. After a considerable struggle, we eventually succeeded. The chromosomes were counted, my report was written. Having shared such an ordeal with them, I gained their respect and when I visited again a few years later they were eager to introduce me to their new giant armadillo, an even more remarkable specimen than the first inasmuch as it was nursing two infants. My joy, however, was soon dampened when, upon seeing the animal, I realized that animal dealers had perpetrated a terrible hoax on my unsuspecting friends at the zoo. The so-called "babies" were obviously not youngsters, but rather full-grown naked-tailed armadillos that had been sold at a high price, I imagine, as baby giants. Worse yet, they were being exhibited to the public as such. The "true" story—I use the word apologetically—clearly demonstrates that we know so little about armadillos that proper identification even by experts often proves difficult.

Today some 20 species of armadillos remain. Yet for some of these, it seems, the final hour is at hand. They are prized as food by man and jaguar alike. In South America their shells are used to make musical instruments such as the Argentinian charango—the bigger the shell, the deeper the bass. Many an armadillo is run over by cars each year on Texas highways, and sadly, some are even made into shopping baskets, tail tied to mouth, and sold in curio shops. Yet remarkably, some, such as the three-banded armadillo, have managed to adapt themselves, forming a very thick shell that can be rolled up into a tight, virtually impenetrable ball at the first sign of danger. Neither man nor the powerful jaws of the jaguar can open it. When the threat is gone—perhaps after an hour—you may see the armadillo trotting awkwardly away to resume grub hunting. Still, however, nothing protects them from a bullet or from drowning as a result of dam construction in their habitats, although holding their breath allows certain species to walk along river bottoms for some distances.

Of all armadillos, the tiny mouse armadillos are among the scarcest. Why they are so scarce no one truly knows. There are two species of this armadillo. One, *Chlamyphorus retusus*, has a total length of 15 cm and weighs hardly more than 100 mg. The back is constructed of a delicate but firm yellow shell, while the soft underside has abundant long white hair. Its small tail, mounted *over* a shell plate, is clearly intended to repel attacks from behind, as one can hardly imagine this mole-like creature easily turning around in the tiny holes it digs through the earth with its disproportionately long front claws and rather unbecoming (for an edentate) shovel-like nose. The other species, *Chlamyphorus truncatus*, is named for its "cut off" mantle. It is generally less hairy, slightly pink, and even smaller than the *retusus*. Both species are endangered. Exhibit specimens of mouse armadillos came to light largely in the present century during construction projects in the few countries that are their home—Paraguay, Bolivia, and northernmost Argentina. Visitors to my office usually tell me that the mounted specimen of *Chlamyphorus* (*Burmeisteria*) *retusus* that I keep there, and that was given to me as a gift by the director of Bolivia's Santa Cruz Zoo, must be a fake of some kind because they find it so difficult to believe that such a creature could exist.

Aside from their physical characteristics and some of their basic habits, not much more is known about the mouse armadillos. Primarily, this can be attributed to the fact that they are very difficult to study inasmuch as they don't venture out above ground as do their bigger relatives. Also, it has generally been very difficult to keep them in captivity for any length of time. Specimens usually last a few months at most. How can we check on their well being if they are mostly underground? What should we feed them? How many cubic feet of earth does a mouse armadillo need for its tunnelling? How do they reproduce? Indeed, how do they find each other in order to reproduce? We have many such questions but very few answers, and little time is left before we will be forced to merely puzzle in front of a museum exhibit over what their life must have been like.

Yet the armadillo may hold the key to at least two very poorly understood human biological phenomena. Two species, the nine-banded and the seven-banded, possess a globoid, single uterus like that found chiefly in primates. Most other mammals have a Y-shaped bicornuate uterus. These two species of armadillos are the only known mammals that regularly produce identical (monozygotic) multiple offspring. In the case of the nine-banded armadillo, which is found as far north as Texas and Florida and as far south as Uruguay, the single fertilized egg splits, after a 3-month period of quiescence, into identical quadruplets, four males or four females. There is never an admixture, cojoining, or any other anomaly as sometimes happens in the offspring of other litter-producing mammals, including man. The seven-banded mulita of Uruguay and Argentina splits the zygote even more often, producing as many as 15 identical offspring. Why and how?

I will close my tale with the other more recently discovered and tremendously significant biologic trait of these underground mammals. Until recently scientists had been unable to find a way to culture human leprosy bacilli for study, with an eye of course toward an eventual cure, as had already been done with the tubercle bacillus. In Louisiana, quite by accident, the armadillo's susceptibility to the bacillus was discovered. When one experimentally infects armadillos with human leprosy bacilli, they rapidly develop generalized leprosy, one of the most dreaded and one of the most untreatable scourges of mankind. Currently 10 million people suffer from this incurable disease, and we can do little more for them than to hide them from the view of society in leprosaria. As all other animals had proved refractory to the infection, there was little hope of progress until the susceptibility of the armadillo was discovered. What else might we learn from any one of the species of armadillos if we studied them more intensively?

The message is simple: We should not allow them to pass with time and certainly we would be foolish to hasten their departure by the wanton destruction of their living quarters. Remember, extinction is forever.

The pronunciation of Przewalski's horse is one of those challenges that one faces when one works in zoos. And, were it not for captive breeding in zoos, this species would now be as extinct as the Dodo, for the last Przewalski's horse was sighted in the 1960s and despite expeditions to seek additional specimens in and around the Gobi Desert, none have been found. It is all the more remarkable then, that from a "founder" population of nine animals, there now exists a captive population of some 700, and plans are currently being made to reintroduce them to their original habitat, thus realizing the conservationist's ultimate dream.

When a Polish Colonel, N. Przewalski, serving in the Russian Army, returned to Europe with a description of this animal and the hide of a specimen shot in Mongolia, an astute zoologist, Poliakov, at once recognized that this hide was not from a domestic horse. He denominated a new species, *Equus przewalskii* Poliakov, 1881. This original specimen is now mounted in a museum in Leningrad. Soon after its discovery, several parties were sent to Mongolia to capture specimens for exhibition and breeding.

Mongolian Wild Horse

(*Equus przewalskii*)

Now one must imagine the difficulties attending such a job. The distances to be covered were formidable and knowledge of the animal, the region, and its people was scant. Moreover, in an age before motorized vehicles were available, how would one race after and capture an animal that was already known to be swifter and much more elusive than the best of Mongolian domestic horses? Because of such difficulties it was decided to capture only recently foaled young who could not sustain the gallop of a chase by horse. These foals were then brought to camp for nursing on domestic horses and, when they were sufficiently grown, marched perhaps 1,000 km to the nearest railroad station for transport west. A number of animals went to the estate of the Duke of Bedford in Woburn, to zoos in West Germany, and to the fabulous Ukranian estate of Friedrich van Falz-Fein, Askania Nova in Southern Russia. Luckily, because of the zoologic interest accorded this animal and the foresight of zoo directors, a reasonably good record of breeding was established, enabling us today to trace most ancestry of current stock.

Przewalski's horse differs in many ways from domestic horses. It is much stockier, and, despite its long captive management, it has remained quite "wild," like the zebra. Rarely has it been tamed enough to be saddled and ridden. To the novice its remarkably erect mane is its outstanding feature, as is its white nose and white "butterfly" stomach. Then, in contrast to all domestic horses with 64 chromosomes (including the Mongolian domestic horse), the wild Mongolian horse has 66 chromosomes.

How it relates to the domestic horse, let alone to other equines (the donkeys, zebras, and hemiones—"half-asses" such as the onager, the kiang, and the djiggetai), is a fascinating but still much debated question. Little doubt exists that at one time its range was a much wider one. Supporting this theory are the colored drawings in the French cave of Lascaux dating back to the last ice age (20,000–30,000 years ago) showing animals whose resemblance to Mongolian horses is striking. We also believe that stone-age Europeans captured wild horses for food by driving them over cliffs. But were these horses the ancestors of our current domestic horse? Or were they, as some believe, the now extinct Tarpan, a mouse-gray horse that at one time lived in the large forests of Poland and Russia? There is no easy answer to these questions and to many others concerning the precise descendency of animals. We know that horse-like creatures began their lives in Northern America with the E*ohippus* some 60 million years ago. Although E*ohippus* was not very horse-like by current standards (it was very small), a continuous line of fossil bones links it to today's *equidae*. They moved over the Bering landbridge to Asia, leaving no relatives behind. Current American horses are all descendants of horses imported since Columbus' discovery of the continent. The so-called "wild" horses of the west, the mustangs, are "feral" horses, runaways of domestic stock, and not truly "wild." Once reaching Asia, so we believe, the horse ancestors rapidly spread through that continent and Europe and, presumably because Asia was then connected to Africa, they also spread through this new continent and "speciated," turning into zebras and donkeys.

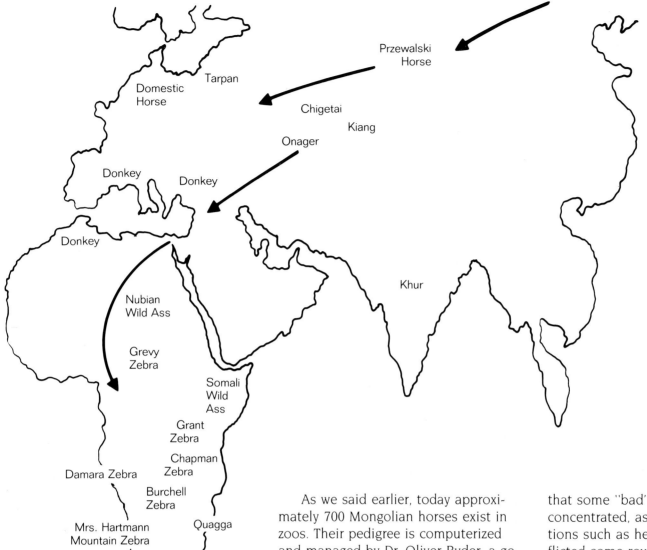

(Above) Distribution of horses and their derivatives over Asia, Europe, and Africa.

As we said earlier, today approximately 700 Mongolian horses exist in zoos. Their pedigree is computerized and managed by Dr. Oliver Ryder, a geneticist at the San Diego Zoo. Like most curators he is concerned with the resultant genetic drift that inbreeding fosters. Coming from so few "originals" it is likely that only some of the genes brought into captivity will survive. Others, unfortunately, will be lost, since not all genes are necessarily passed on to each offspring. It is equally possible that some "bad" genes may become concentrated, as happens with conditions such as hemophilia, which has afflicted some royal families in Europe. Moreover, knowing so little about exactly what the genetic make-up of the original horse was, we fear that zoo breeding may inadvertently select unrepresentative genotypes. For instance, having no harsh winter in San Diego, our environment may select animals that produce much less winter fur, or teeth no longer suited to the harsher food of the wild. As a first measure to combat such selection we chose to exchange animals not only among conventional zoos, but also with the stock in Askania Nova, which has grown separate from animals in the west for decades. In 1982, following extensive negotiations, three animals from the United States went to Russia, and three came from there to us.

Curiously enough, one of the animals we received, a mare named Vata, was immediately diagnosed as pregnant. We congratulated ourselves on what we considered to be a fine bit of horse trading, getting four for three. And Vata was well received at the San Diego Wild Animal Park by Basil, the reigning stallion of our herd. But as soon as the foal was born and on its feet, Basil, in a display of uncharacteristic behavior, fiercely attacked the youngster, nearly killing it instantly. The attentive keeper, Rick Massena, tried to rescue the infant by pulling it onto his truck—only to be attacked himself, his arm getting badly crushed. But his efforts were all for naught as the foal, Vargo, died anyway. Still in "foal heat" though, Vata was promptly served by Basil and 11 months later she delivered Vasiliy, under the now caring eyes of Basil. The inference is, of course, that Basil "knew" that the Russian mare's foal was not his own. But how did he know? Could he smell it? Did he know when Vata arrived that she was pregnant by someone else? Surely he cannot count. Just how is it that he felt motivated to exclude this youngster from his stock?

These are challenging questions for behaviorists, important observations for zoo managers as well, and perhaps testimony enough to convince the skeptic that the current zoo stock must still be considered "wild." And that is, of course, the challenge to modern zoos, to preserve animals in their original state. And not just this one, but all species in their care. You can see from all this that going to visit a zoo can become a very meaningful experience. Our appreciation of wild animals can take many forms, but if we endeavor to save them we will require a lot more knowledge than we currently possess.

Where do we go from here in saving the ancestor—so I believe—of our beloved domestic horse? If you think SALT negotiations are difficult, try establishing a preserve for Przewalski's horse in the Republic of Mongolia! Of course zoos will donate representative animals and pay the price for shipping them, however much the cost. But consider the adaptation problems attending taking the animals from balmy San Diego (minimum temperature 0°C) to Mongolia (−20°C). Do you think that the zoo specimens would find food and water right off the bat in the Gobi Desert? Do you think veterinarians in Mongolia will be up to the task of supervision until true adaptation to the wild has occurred? And what about fencing material for the initial stock? Will American airlines fly it in to be transported on roads that have yet to be built? Will it be erected by Mongolian fencemakers who at the moment don't exist? Or, for that matter, will snow drifts obviate the construction of fences altogether?

These are some tough questions that need concrete answers. In them you can see the practical foundation of conservation laid bare. But succeed we will, because the zoo community is committed to preserving these beautiful wild horses for all to see and enjoy in the future.

Uniquely American animals, the peccaries do not enjoy the reputation of being particularly beautiful. And yet, anyone who has ever encountered the spectacularly colored African red river hog or witnessed a mother warthog being followed through the savannah by her offspring, their tiny tails erect, knows that despite their reputation, pigs can be beautiful.

    The Suina, as the larger taxonomic order is named, had their origin in eastern Asia. From there they dispersed and speciated into true pigs, the Suidae of Europe and Africa. However, those animals that came over the Bering landbridge from Asia to America millions of years ago changed drastically. They became the ancestors of today's peccaries, or Tayassuidae (*tayacu* being Brazilian for peccary). True pigs never existed in America. The wild boars now living in the hills of Tennessee are descendants of animals introduced by wealthy landlords for sport

Giant Chaco Peccary

(*Catagonus wagneri*)

hunting. The domestic pig, *Sus scrofa*, was also introduced by American settlers. Before that time there existed only peccaries, and before North America was connected to South America, peccaries existed only in the northern hemisphere. At about that time (perhaps some 8 million years ago) the "Great American Interchange" (as the late paleontologist George Gaylord Simpson aptly called it) of animals and people began to take place. Thus, animals such as the porcupine, the opossum, and the armadillo moved north as far as their ability to survive in the new environment would permit. And rather more animals moved south during this remarkable period. All the South American deer and its very endangered spectacled bear are derived from North American ancestors, as are the cats (jaguar, ocelot), dogs (bush dog, maned wolf), and of course, the peccaries.

Peccaries differ from pigs in many ways. For instance, true pigs have upturned canine teeth in their upper jaw while those of peccaries are directed downward. The scent gland of peccaries, located on their back, that earned them the designation *Nabelschwein* ("navel pigs") is absent in true pigs. The construction of stomachs, hooves, and many other anatomical structures differs as well. Peccary bristles, as opposed to the solid hair shafts of most mammals (including man), are honeycombed in their interior and must provide wonderful insulation to the animal.

Until recently it was thought that there were only two species of peccary in existence: the collared peccary, which ranges from Texas all the way to Argentina, and the larger white-lipped peccary, which ranges from southern Mexico to Argentina as well. And this was the state of affairs until barely a decade ago. These two species were not endangered, and little attention was paid by science or the public to peccaries as there seemed to be an unending supply. Then, a third species was discovered by the late Ralph Wetzel in the Gran Chaco of Paraguay. Wetzel and his colleagues named it *Catagonus wagneri*. Unfortunately however, we must report that no sooner has this animal been discovered than it is perilously close to extinction.

The Chaco, or "green hell" as it has often been called, is a remarkable region. At the time when the Andes mountains rose from the South American continent, their earthen cover washed off to fill a giant trough to the east. Although parts of this trough are in eastern Bolivia and northern Argentina, most of it is now in western Paraguay. *Chacu* is the Indian word for the "plains of animals." And this vast region, virtually flat and stoneless, is just that. It is rich in fertile sandy soils, and full of unusual trees, bushes, and animals. It is the home of the

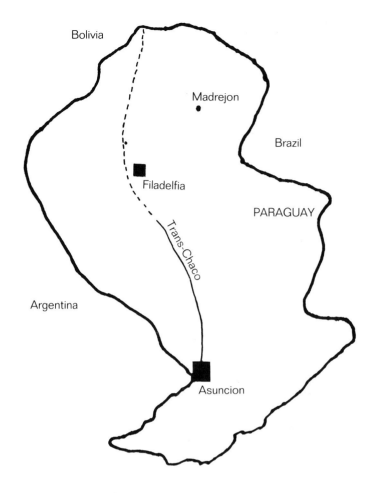

(Above) Map of Paraguay with Trans-Chaco Highway, the Mennonite colony of Filadelfia, and the center of the national park at Madrejon.

quebracho (Spanish for "ax breaker") tree whose 15 percent tannin content has been such an economic boon to Paraguay.

And Paraguay, with its nearly 3.5 million inhabitants, harbors many such treasures and surprises. Fiercely nationalistic, the Paraguayans fought two large wars with their neighbors and depleted much of their male population. The last war, fought against Bolivia in 1932–35 over oil deposits that were never found, fostered the construction of the Trans Chaco Highway which, through enormous effort, was cut from Asuncion, Paraguay's capital, to the Bolivian border, some 435 miles to the northwest. Nearly straight as an arrow, this clay road leads through the thorny, virtually impenetrable bush. When eventually the final battle in the war was fought, and no oil was found, slumber fell once again over the region, and the Chaco was again turned over to its principal inhabitants: mosquitoes, jaguars, and jacares, the local crocodiles. Turned over that is, with one important exception, for, much to the surprise of the Paraguayan soldiers, they discovered a flourishing Mennonite colony deep in the Chaco! In three more or less distinct periods of immigration beginning in 1926, Mennonites from America, Russia, and Canada had built up a prosperous farming community of about 10,000 and a bustling center of activity which they dubbed Filadelfia. Here blue-eyed blond kids learn an older German dialect as their first language, then Spanish, and then Guarani, an expressive Indian tongue spoken by nearly all Paraguayans. The members of the colony lead an impressive communal agricultural life. They are surrounded by several formerly hostile Indian tribes that have now settled in permanent villages, and who communicate with each other and the Mennonites in, of course, German! The Mennonites and the Indians had known the Giant Chaco Peccary long before science did. Indeed, it is their preferred wild meat, as it smells little, is very lean, and it is easy to hunt. And therein lies its tragedy. For of all the species of peccary that are found in the Chaco, the giant peccary is perhaps the "dumbest." It travels in small family groups, perhaps 5 to 10 animals. When one is shot, rather than scattering (as do the other species) they investigate what has happened to the fallen animal and so become targets themselves. Thus their population has dwindled as the use of firearms has become more widespread, as motorbikes allow more rapid access to the interior, and especially as land is cleared for cattle farming, jojoba ranching, and the like. Over recent years, some 400 km of the previously clay highway have been asphalted and every year another 28 km of tar is being laid. Weather conditions subsequently no longer impede cattle transport and soon a journey to Bolivia will be as feasible as a trip to Filadelfia already is. Heavy equipment such as tractors and "rollers" can now be brought in to wreak havoc on the formerly impenetrable bush. Everywhere one sees pillars of smoke from enormous fires made from cut-down brush. The Chaco peccary exists only here in the Chaco which, it seems, will soon vanish. What will we lose and what can be done?

If we allowed the giant Chaco peccary to be lost to extinction it would be tragic for many reasons. It is the largest newly discovered mammal and consequently is regarded as the least studied and most poorly understood. We know little of its general biology. We have however learned that it usually has two offspring which in former times were born in January. But now they farrow between August and December. Why? Has the climate changed? Are the numerous tajamars (waterholes for cattle) responsible? We have literally only begun to gain the slightest insight into its lifestyle. Then there is the matter of its genetic relationship to the other species. Dr. Wetzel considered it to be descended from or closely related to an extinct form. Our own chromosome studies, on the other hand, show that its sparse 20 chromosomes differ from the 30 and 26 of the other two species. Furthermore it possesses the most unusual stomach structure of all the peccaries, adapted as it is to living mainly on prickly pear cactus and the like. Numerous other reasons could be given why it would be sad, if not tragic, if we lost this large animal without even trying to save it.

Yet only one male specimen lives in the Asuncion Zoo. Captive breeding seems as elusive as it is *in situ*. It is not an easy task to bring such animals into a zoo for propagation and study. Adult animals are virtually uncatchable. There are no traps for them and their peripatetic habits in the bush makes locating them nearly impossible. And even if they were caught, exportation would be virtually impossible due to the special permits required and also because of genuine health concerns. So there remains only *in situ* conservation which is clearly the best way to undertake efforts directed at the preservation of such a wild species. The Paraguayan government and its conservation agencies are thus to be congratulated for having set aside a 250,000 ha area of the northern Chaco as a reserve for most of the country's unusual flora and fauna. When I last visited there I was able to catch a young male "tagua" and bring it to the female already there at Madrejon, the area headquarters. With some luck a colony will be started in this park and scientists will hopefully be able to enjoy some of the most spectacular but least understood mammals of South America.

Even the more erudite of my friends shake their heads when asked about the Okapi. "Never heard of it" or "it can't be important," they usually say. But what a wondrous beast it is. Could you imagine purple as a color for *any* animal? The Okapi is deep purple in color and has zebra stripes to boot! But it isn't a zebra. We know that from its hooves which are double toed as opposed to the single hoof of the equine family. So what are they? Just imagine yourself a short 80 years ago exploring the bush of Zaire—then the Belgian Congo—when Sir H.H. Johnston, governor of Uganda, learned of this big animal which had never before been described. Because of its size and its stripes it was, at least at first, considered a forest zebra. When Johnston presented Sclater (the curator of the Capetown Museum and authority on African fauna) with two belts prepared from the hide of an Okapi, which he had obtained from natives, scientific curiosity was aroused. The Okapi is now known to weigh 400 pounds. So how could it ever be overlooked for all those years? Early observers probably considered it to be some type of zebra hybrid because of its markings. But it is a "good species," with no known immediate relatives.

Okapi

(*Okapia johnstoni*)

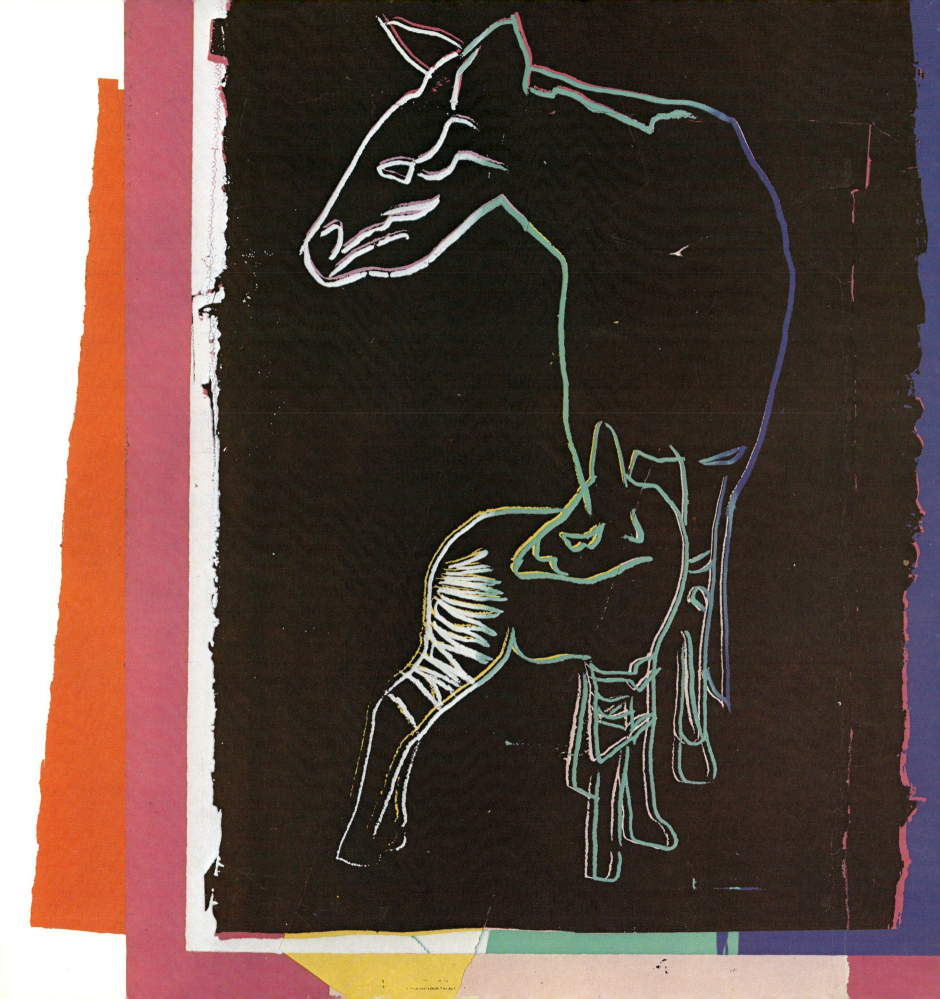

In 1901 when the first hide and skull arrived in London the former "*Equus johnstoni*" was quickly renamed for various complex reasons, not all necessarily logical. But since hooves were attached to the hide it was correctly placed among the artiodactyls into the family *Giraffidae*. And the okapi is indeed an artiodactyl, or "cloven-hoofed" animal with some similarities to giraffes (its long neck) and other ungulates like the nilgai (its pellet-shaped feces). The name "okapi" was given to them by the pygmies of the dense Ituri forest of the Congo. Since okapis are secretive animals, their lonely habits are poorly studied and—until the discovery of the giant Chaco peccary in 1972—it was known to zoologists mostly as "the last large mammal discovered." Few of us have had the privilege of seeing it "close up and personal."

Imagine the initial difficulties of bringing such an animal to any zoo! But the Belgians were determined. The first two animals to arrive in Europe did not survive very long but the third, arriving in Antwerp in 1928, lived for 12 years. A station with an airstrip was eventually constructed in Epulu, northern Zaire, and since 1949 the animals have been flown out of the jungle with more success than that which attended their initial long sea voyages. Imagine carrying a 400-pound animal through miles of forest to the station at Epulu! Their captive population has been around 60 during the last 15 years and of the 21 held in America, five live at the San Diego Wild Animal Park. The

(Left) Young okapi sucking while mother cleans rear. (R. Garrison, San Diego Zoo)

43

curators of the venerable Antwerp Zoo are, not surprisingly, the keepers of the okapi studbook. A studbook, not unlike those used by horse breeders, is kept for all severely endangered animals. And they have served good purposes as we will see. All captive okapis are further managed by a species survival plan (SSP), a committee whose objective is to "increase numbers in captivity while developing a more even Founder representation and minimizing inbreeding."

The okapi is a browser, not a grazer like the cow. It extends its neck and sticky blue prehensile tongue to reach for leafy morsels from many types of trees. You might well ask then—what do you feed them in Antwerp let alone in the Brookfield Zoo of Chicago? This is obviously a problem for zoos, not only in their care of okapis but for most of their precious charges from foreign lands. Indeed you may ask, is it justified then to keep okapis in zoos—far away from their natural environs? The answer to such a question will be, no doubt, a very personal one. "No" say some, very emphatically; "yes" say we who have witnessed the fate of Przewalski's horse and other species that would be extinct were it not for captive breeding. But if you are truly objective when observing the okapis at San Diego's Wild Animal Park, you will have to admit that their lives are pretty comfortable, if not natural.

And the okapi has, of course, some characteristics that are in need of further study. The genetics of their striped hindquarters, for instance, is so variable as to allow a photograph to be its tatoo. Yes indeed, we tatoo precious zoo species, thereby avoiding mischievous trades, allowing studbook entries, assuring proper pairing, and giving veterinarians confidence that they are treating the right animal. In okapis a look at the hindquarters suffices for identification.

And speaking of hindquarters allows me to relate an embarrassing yet important story. Clearly the successful breeding and rearing of young okapis is of paramount importance to enlightened zoos. As a matter of fact the larger zoo community would not permit flagrant mismanagement of *any* species for very long, least of all its most precious members. But the rearing of okapi young posed some significant challenges to curators in the past. There was, for instance, the question of how long it was wise to wait for a calf to discharge its first stool, or meconium, before veterinary intercession would be called for. When young in several zoos were closely observed it was thought that this might take several weeks. Is that "normal" in the wild? We couldn't believe it. So, to learn what okapis do in the wild became an important question. To the best of our knowledge they deposit their young in hiding while they forage and nurse at intervals. But there are no infant-hiding places in zoos and since food is supplied (with the best of intentions), there is no need to browse. So what does the mother do with her spare time? Cuddle the infant of course, and lick its rectal area. The result is rectal prolapse, which is an eventration of the terminal bowel with ensuing infection. It frequently resulted in death. As recently as 10 years ago we puzzled over the peculiar okapi bowel anatomy that seemed to promote this condition and over the delay of passing meconium that frequently necessitated laxative

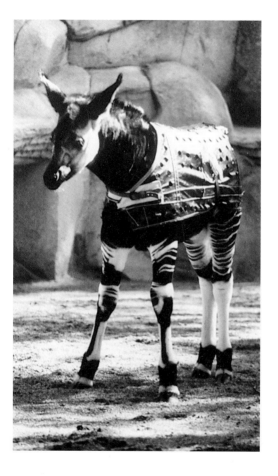

(Above) Harness made for infant okapi to prevent rectal injury by mother's tongue. (R. Garrison, San Diego Zoo)

treatment. How innocent we were and how simple was the solution. Avoiding boredom is the somewhat simplified answer, but there is more to it than that. In our zoo we went so far as to construct harnesses that would curb the maternal behavior of licking the rear end of its young. Because of boredom—so our behaviorist concluded—the mother (in captivity) much extends the duration of an otherwise brief and less frequent behavior. With her long harsh tongue her prolonged licking would in due time cause the prolapse.

Why would anal licking occur at all? We conjecture that it is done for the following reason. Most animals are born without the immune defense system that develops later in life and which then provides vital protection against bacteria. Only gradually do we develop antibodies and secrete them into our bloodstream. It takes from weeks to months for lymphocytes—the circulating white blood cells in charge of making these antibodies—to become "trained" to function adequately.

Mother Nature thus developed a neat mechanism to help out—mother's milk supplies the antibodies and, for a day or so, the baby okapi's stomach lets these antibodies pass into its blood. Thus, the first defense against the worst offenders is guaranteed. The first milk (or colostrum) is therefore a vital nutrient to any calf. But that may not be all. Zoos, as well as farmers, at times have the difficult task of hand-rearing animals while mother's milk is unavailable. We have used cow's milk in San Diego, as well as other substitutes borrowed from human infant nurseries, with careful attention to fat, sugar, and protein content—only to find that despite all this detailed care many youngsters suffer, and even succumb, to intractable diarrhea. No antibodies help and unusual organisms, for instance Cryptosporidia (the cause of many a traveler's diarrhea), are the cause. A healthy adult cow (or a traveler) will ultimately come to grips with the infection. They make antibodies with lymphocytes that reside in large numbers in the gut surfaces. A calf does not yet have this protection. It relies, so our present "wisdom" tells us, on its mother's milk. By sampling the infant's anus with her tongue, perhaps once a day in nature, she perceives with her own immune system the harmful intestinal flora that upsets the baby's digestion, quickly produces antibodies (her system is already trained to do so), and delivers these with the next day's milk to her young—in time for it to recover from the intestinal infection.

Having at last made these circuitous connections in our minds, we now feed preimmunized cow's milk to all our handreared gazelles with the most magnificent results. Of course we no longer allow boredom for mother okapis either, and the harness has now become a part of ancient history. Our attention is now directed to other problems and surely we have grown wiser. Now more often we try to emulate nature when keeping wild animals and zoos have accepted the notion that such research is needed, lest species like the okapi wander off into the mist of extinction.

For a while it looked as though the whooping crane would follow the passenger pigeon into extinction. Decline in its numbers was so alarming that there appeared to be little or no hope. In 1941 only 21 were known to be alive. But an enormous rescue effort began and by 1968 the number had risen to 48, and to 59 in 1971. However in 1973 there was a terrible set back and only 43 birds remained. Once again it was thought that extinction was imminent and conservation efforts were redoubled. Again the whooping crane population rose and today there are 96 known to be living in the wild. What is more important, there are now two distinct populations in the wild and this is in addition to a captive flock. We now know something about their biology and we can afford to be a little more optimistic about their future though they are by no means "out of the woods" yet.

Whooping Crane

(Grus americana)

Cranes belong to the family *Gruidae* (*Grus* is Latin for crane) and are among some of the largest and most beautiful birds in existence. They stand more than four feet high and can attain a wing span of up to 7½ feet. While they might look like storks, they are not related to them. An important difference, for instance, is that while young storks are nestbound and helpless, cranes leave the nest on the first day. Likewise, their similarity to herons is only superficial, as cranes stretch out their necks during flight, while the herons' neck is curved backwards in a characteristic S-shape. Moreover cranes do not like to sit in trees, and only one, the African crowned crane, will build its nest off the ground.

There are 14 species of cranes, five of which are already very rare or near extinction. Some problems that may impede their survival are that they steadfastly stick to the same breeding grounds, they employ the same flyways to the same wintering grounds each year, and they pair-bond for life—small wonder that they are considered the symbol of fidelity in Oriental culture. Add to that the fact that they make loud trumpeting calls when courting and during flight and it is easy to understand why they are easy prey, particularly for man. The trumpeting sound from which our "whooper" takes its name is produced in the true cranes by an elongated, spiraled trachea, half of which (in whooping cranes) is tucked away in special hollows in its sternum. This unusual windpipe comprises over 300 bony rings and is a conspicuous part of the generally unusual anatomy of this bird.

Whooping cranes are long-legged, large white birds with black primary wing feathers and they have bare, reddish skin around their bill. The black feathers are hardly visible when the bird is standing. In flight, however, a large portion of the wing tips appears black when seen from below. At times the birds are mistaken for the grayish Canadian sandhill crane, particularly since they may mingle in migration. Indeed it is thought that conservation activities were long delayed for the whooping crane because the larger numbers of migrating sandhill cranes had been mistaken for whoopers.

Like most other cranes, the whooping crane lays two large eggs in May, having beforehand constructed a large, foot-high nest in the very inaccessible marshy areas of North America. The second egg is laid two or three days after the first and consequently one chick hatches two or three days later, some 32–34 days after being laid. Being slightly older and stronger, the first chick has a decided advantage over its sibling, and since young cranes are known to squabble and fight fiercely, the second chick often dies, being trampled to death or thrown out of the nest. Both parents feed the hatchlings and they grow quite rapidly. Such rapid growth enables the young to take part, only 4 months after birth, in the long flight to the wintering grounds in Texas, some 2100 miles away. When having to go farther north to find nesting grounds, the consequent shorter summer period may not always allow the young birds to reach a level of maturity equal to the longer flight to Texas, and as a result, juvenile mortality has increased. It has been estimated that an 80% first-year mortality takes place among the young. In part, this is also the result of the young being driven away by the parents at the age of about eight months. Perhaps the young cranes are thus encouraged to find a mate of their own. Sexual maturity is reached after 5 years, although the adult plumage is attained by the age of 2. Longevity is probably around 40 years. Thus, once reaching maturity, and given no serious threat to their environment, the cranes potentially have a long and fertile reproductive life ahead of them. Why then has it been so difficult to increase their numbers and why are the scientists of the "Recovery Team" so cautious when they give prognostications for the survival of our largest crane?

In 1945, National Audubon Society specialist Robert P. Allen was appointed to undertake a detailed study of the whooping crane because its extinction was felt to be imminent. It is primarily because of his spearheading efforts that this majestic bird has survived. He aroused wide publicity on behalf of the cranes. But above all Allen set out to learn precisely all about the biology of this large crane, and such efforts continue, even today, at Patuxent Wildlife Research Center in Laurel, Maryland and at the International Crane Foundation in Baraboo, Wisconsin. The most important task initially was to find the nesting grounds of the birds. Their wintering grounds in the gulf coast of Texas had been located as early as 1937, and this area, known as Aransas, was established as a protected refuge. Over a 4-year period, enlisting people from all walks of life, Allen sought their breeding grounds in the far north. Finally, in 1954, quite by accident a helicopter spotted young near the Great Slave Lake in western Canada, a very inaccessible area. In earlier times the birds had been distributed over a much larger range of the marshy areas of North America. A large concentration had existed at the North Platte River in Nebraska. But they disappeared from these nesting areas because the marshes had been drained dry. So they disappeared in the summers from the USA and sought refuge in Wood Buffalo Park, south of the Great Slave Lake on the border between Alberta and the Northwest Territory. The park was strictly protected. In 1966 six to eight breeding pairs were known to exist and by 1985 28 pairs have nested in the park and they laid 21 eggs. Of these, 16 hatchlings were banded and these birds were known to have migrated to the Aransas refuge in Texas in the same winter.

With the discovery of the nesting grounds and the protection afforded to this area the number of cranes gradually increased but then several setbacks occurred. In 1973 scientists were alarmed that of the 59 birds counted in 1971 only 43 were still alive. Snowstorms were known to have caused occasional losses, but such a drastic reduction was frightening and even today remains unexplained. Efforts were then redoubled and the most important steps were taken that eventually led to the creation of two separate flocks of whooping cranes. One group of birds continued to migrate between Wood Buffalo Park and the Aransas refuge, and the second group followed the sandhill crane population to their wintering and nesting grounds.

This "division" of the population into two flocks was no accident. It was decided that inasmuch as the second egg of whooping cranes so often hatches the "weaker" chicks which do not survive, it would perhaps be wise to "steal" them and attempt to raise them by other means. Six were "stolen" in 1967, 10 in 1968, and another 10 in 1969. They were then taken to the Patuxent Wildlife Research Center and hatched in incubators. But they still not only had to be reared successfully to adulthood, but a way had to be found to ensure that the birds would assume appropriate whooping crane behavior. This was a difficult task. The natural conditions of light and environment were simulated. Eventually even the well known and ritualistic dances of premating cranes took place in this flock, but the pairs did not mate.

For the proper management of captive whooping cranes, as was the case with both the Puerto Rican parrot and the California condor, we had to learn how to differentiate the sexes of the birds because both males and females look alike. Currently the chromosome method used for condor sexing is also being employed for cranes. In 1975 it became possible, for the first time, to artificially inseminate female cranes.

Subsequently, in 1978 a female laid 10 fertile eggs following insemination, and soon another laid nine more. Meanwhile crossfostering was deemed a possible way to produce a larger number of birds. Thus, in 1975, 14 eggs were transferred by helicopter from Wood Buffalo Park to nesting sandhill cranes at Grays Lake in Idaho. Of the nine that eventually hatched one died in a snowstorm and two simply disappeared. But five migrated with the sandhill cranes to their wintering grounds in the Bosque del Apache National Wildlife Refuge near Albuquerque, New Mexico. Now eggs are being transferred from the artificial insemination program in Patuxent directly to sandhill crane nests. Through all these efforts there were, by 1980, 76 whooping cranes in Wood Buffalo Park, 15 with the sandhill cranes at Grays Lake and 28 in captivity, mostly at Patuxent, all making a grand total of 119 "whoopers!" And, at the Texas wintering grounds, the all-time high of 96 birds was sighted in the winter of 1985–86.

But will the Grays Lake group be "good" whooping cranes or will they assume the ways of the sandhill cranes? They are true to the migration patterns of their adoptive parents, but this was expected. On the other hand though, will they "speak" as whoopers or sandhills and will they perform the ritual dances by their genes or those learned from the sandhills. In other words, will animals from the now separate pools of whooping cranes ever recognize each other as members of the same species if they should mingle in the future? Even worse, might they hybridize with the sandhill cranes? This can be done artificially but would it happen in nature? These are very important questions whose answers will come in the near future. And the answers will dictate the actions of the whooping crane recovery team in the future.

The aim is to have 1,000 whooping cranes flying in the next century, secure from the perils of extinction and divided into several populations. In his 1967 book, *Extinct and Vanishing Birds of the World*, J.C. Greenway wrote that "the survival of the species would be a miracle and that "we may hope, but perhaps we cannot expect that the cranes will survive." It delights me to say that there is now more room for optimism where the whooping crane and a few other crane species are concerned. They are indeed majestic animals and we would be well advised to adopt the oriental reverence for this beautiful group of birds.

The family Varanidae takes its name from the Arabic *ouran*, their word for the gray desert monitor. Varanidae is a relatively small but colorful family of dinosaur descendants. Its members are often referred to as "dragons" and the Komodo dragon is the largest of the 42 monitors in existence. And yet it was not until 1912 that it was first known to the scientific world. The director of Java's Buitenzong Botanical Garden, Major P.A. Ouwens, had heard rumors of the existence of "land crocodiles," as the animals are referred to locally, and he eventually secured some of the first specimens. He named and described them and so spurred on the scientific and zoological community all over the world to acquire specimens as well. Fortunately the Indonesian government has been very careful about releasing many of the animals. Consequently the species is surviving well on the four islands that comprise the Komodo National Park, some 350 miles east of Java.

Komodo Monitor

(*Varanus komodoensis*)

The Komodo monitor is the largest and most primitive living lizard. Its yellow forked tongue is perhaps its most memorable feature when one first encounters it. It looks ferocious, and, when lifting its body high off the ground with its powerful legs, it is capable of achieving speeds of up to 30 km/h in pursuit of its prey. It has a gray-brown color with a lighter underside and a long powerful tail that it uses to knock down large animals which it then attacks with an equally powerful jaw. Its half-inch teeth constantly replace themselves. The teeth are sickle-shaped and their rear curve is serrated, and thus they are ideally suited not only to sinking into and holding onto prey, but also to shearing the meat apart. Each of the monitor's feet has five digits and is endowed with long, sharp claws that further enhance the fierceness of this carnivorous beast. The skin has a fine granular scaly texture that shows rather less pronounced rosettes than are seen in most other monitors. They also frequently have stripes and dots, and may bear many different and striking colors or combinations of colors including blue, green, yellow, and brown.

Not many people have been to the Komodo Islands. The approach by boat is treacherous, and yet it is the only way to reach them. Some 500 fishermen and their families live in the only town on Komodo Island, Kampung. Here a tourist facility receives 500 or so visitors that now come to the island annually. The climate is very hot, and the dragons live in dug-out earthen holes or under brush to protect themselves from the heat. This is because lizards are ectomorphs, meaning their temperature is largely dependent on the temperature of their environment. And since the daytime temperature on the islands may reach 110°F, lizards (particularly the young) must be careful as they are known to die if their body temperature reaches 108°F. And, unlike mammals, they cannot control their own temperature very well at all.

Monitors are diurnal animals, stirring at 4 a.m. and climbing out of their burrows soon thereafter. They then wait for the sun to warm them and by 9 o'clock they are ready to stalk their prey which consists mostly of large animals. Rusa deer are one of their favorite targets but they also like pigs, water buffalo, goats, dogs, horses, and just about whatever else might come along. Younger animals eat crickets, birds, eggs, geckos, and once again, whatever else they can get. All are carnivorous. However despite their ferocious appearance, most monitors will flee man with a hiss, the only sound they are capable of making. Only on rare occasions have they attacked people. Indeed, a famous Komodo monitor at the Bronx Zoo in New York was so benign that it could walk without a leash among the zoo's visitors.

Komodo monitors frequent the same paths every day and here they lie in wait for their prey. In fact, most of their life is spent in waiting. Mostly they wait for deer herds to change locations, which also takes place on predictable paths. And when one of the deer comes within striking distance—

(Below) Shed skin of Komodo monitor showing the structure of its tiny scales.

about three or four feet—the powerful jaws sink into its body. The victim is then wrenched rapidly to the ground, and as quickly as possible the abdomen is opened and the internal organs devoured. Nearly all the prey is swallowed in big gulps, bones, feet—everything. But, if by chance the prey escapes after being bitten, so many vicious bacteria have been injected with the monitor's saliva that the poor animal will soon die from blood poisoning anyway. Of course a meal of this size lasts for awhile. Indeed, captive monitors get along fine on two pigeons a week or the equivalent. Monitors are also thought to eat each other or at least adults may devour young animals, although this aspect of their behavior is poorly documented. To be sure they will eat a dead monitor, but whether the arboreal habits of youngsters are a response to the fear of being eaten by their elders seems less probable. More likely they are attracted to these haunts because this is where they find the birds, geckos, and other bite-sized animals that are their prey.

Young Komodo dragons are about 10 inches long and weigh about 2½ ounces at birth. They are hatched from eggs after an 8-month incubation. We believe, as with the tortoise, that gender is determined in response to specific environmental temperatures at critical stages of development. How-

(Below) Komodo dragons at San Diego Zoo, the male extending his forked tongue. He is shedding his skin in flakes, a characteristic of normal growth. Note how high they lift themselves off the ground. (R. Garrison, San Diego Zoo)

ever, what this temperature is remains, as of yet, unknown.

Although quite a few Komodo monitors have reached many zoos in the 70 years since their discovery, only in Surabaja, Indonesia have any fertile eggs been hatched. What could be the problem in other zoos? It is just possible that pair-bonding occurs in these animals. Reptile researcher Auffenberg spent a year with the dragons in 1969 and encountered courtships and mating from July to October. He had the definite impression that pair-bonding was present. Moreover he estimated that there were only about 300 sexually mature animals on the island. He also observed that eversion of the hemipenises (male reproductive organs) induced intense tongue flicking by females and he conjectures that possibly pheromonal attraction occurs before mating. Few monitors have bred in zoos and so far no varanid lizard has had a second generation in captivity. Could it be that other factors in captive management are awry? Physiologist John Phillips of San Diego suspects that the animals are not kept warm enough. In comparing the Djakarta Zoo animals with those in San Diego he noted that the Djakarta animals were much more active and feeding seven times more than their San Diego counterparts. When he measured their intestinal temperature he found it to be 37°C in Djakarta, while only 34°C in San Diego. And since it is well known that reptilian animals regulate their functions by external temperature it seems likely that, in colder climates, their reproductive system shuts down. Their energy metabolism is probably busy just keeping them warm. So Phillips constructed aluminum water beds and found that the Komodo monitors, as well as many other reptiles, tried to seek warmth when offered a choice. Perhaps this is the reason why these monitors love to lie in pools of warm water which, incidentally, may also free them of ticks.

So what are the dangers to the Komodo monitors? A second town has been settled on the islands in recent years and it is possible that human population pressure may expand even further. Tourism is being strongly encouraged by the government and most likely this will lead to some disturbance of the dragons' way of life. At present the Indonesian population of the islands consists only of fishermen. But with a burgeoning population it is possible that deer hunting may rob the monitors of their principal food source. The fishermen now live in huts built on high stilts and they do not hunt the deer and other mammals on which the monitors feed, but the clearing of land to make way for new housing and increased tourism may become future threats to their existence. It is also feared that mineral deposits may be discovered on the island, a prospect that would alter the situation at once.

But however much the Western zoological community may dream of having additional monitors for exhibition, until such time that their requirements are better understood, and they are able to successfully reproduce in captivity, it is best if this prehistoric beast is left protected on the islands to which it is native.

Only with some trepidation did I decide to include the Sumatran rhinoceros in this book. After all, there is no zoo in which the reader might see one. The last captive specimen, a female, died in the Copenhagen Zoo in 1972. And perhaps as few as 100 of these prehistoric looking animals are suspected to be left in the five widely separated areas of Sumatra that are its home. For this reason an intense international effort is presently afoot that is endeavoring to collect a founding stock for captive breeding in carefully selected zoos. But the story of this elusive animal, once widely distributed over Sumatra, Borneo, Laos, Vietnam, and Thailand, will allow me to expose the problems being experienced by all five rhinoceros species.

Sumatra's Rhinoceros

(*Dicerorhinus sumatrensis*)

The main problems of rhinos are that they are large and that they possess horns. Their size requires that they have a lot of space in nature, and, because they have relatively inferior eyesight, they are easy targets for today's high-powered rifles. It is difficult to understand, however, why rhino horn, which consists of matted hair-like material, should be such a desirable substance to man. Contrary to popular belief, it is rarely sold as an aphrodisiac. It is most desired as a medicine against febrile illnesses in Oriental culture. There one may obtain powdered rhino horn, carefully weighed out in apothecaries, for $300 an ounce and up. But the most rapidly escalating consumption of horns has come from the recently oil-affluent Yemeni, who find rhino horn the macho "in" thing from which to fashion elaborately carved sheaths for their daggers. All this has led to an alarming decrease in all rhinoceros species.

In Africa two species occur: the solitary and seemingly mean black rhinoceros, *Diceros bicornis*, whose population is down to 6,000 from 13,000 in 1980, and the white or "square lipped" rhino, *Ceratotherium simum*. When Ian Player, then chief conservator of Natal, decided on a conservation/relocation action because of the severe pressure being experienced by this species, we imported 20 animals to our San Diego Wild Animal Park in 1972. From these we have raised 55 offspring. It is now the only species that has increased in the wild from about 3,000 in 1980 to about 4,000 at the present time.

(Above) Sumatra's rhinoceros adult female at Copenhagen Zoo (by permission).

The three remaining rhino species are Asiatic. The largest and best known is the Indian rhinoceros (*Rhinoceros unicornis*), which was the first rhino to be exhibited in Europe. It prefers swampy grounds, has what appears to be an "armored" exterior, and it is well managed in India as well as in zoos, with a total population of over 1,000 specimens. Next there is the Javan rhinoceros, *Rhinoceros sondaicus*, so elusive a creature that even few photographs of it exist. The Javan rhino is confined to the peninsula Udjung Kulon, and the best estimates suggest that only about 50 of its kind still exist. Stringent laws against poaching and protection of its habitat are all we can hope for at present. The situation for the third Asiatic species, the Sumatran rhinoceros, *Didermocerus (Dicerorhinus) sumatrensis*, is a little different. Estimates put its surviving numbers at about 300, but one student who recently tried to observe it spent over a year on its trail without ever seeing a single specimen! It lives dispersed in the thick jungles of Sumatra and is extraordinarily difficult to follow and study. Moreover, the trees that comprise its jungles have been sold, destined to become furniture, packing crates, building material, and firewood. Perhaps it is because its habitat is doomed that conservationists recently have become interested in it. Or perhaps it is because it is the only truly woolly species, or perhaps because we feel ashamed of having let the last female captive die unfulfilled. Whatever the reasons for the upsurge in interest, and despite the fact that over 50 of these animals have reached European zoos in years past, the success with keeping Sumatran rhinos in captivity has been dismal. The relatively long survival of the Copenhagen specimen was due in large part to the sandy exhibit space in which it was kept and to the attentive efforts of its keeper. For it was only through his observations that the rhino's preference for alfalfa was discovered. Offerings of all types of food had failed to stimulate her appetite. But when the keeper noticed her intently sniffing the air as some freshly cut alfalfa was being driven by, the barrier was broken. She later specialized in eating loaves of fresh bread and buckets of ripe apples. And once again this gives me the opportunity to tell one more tale of the past—dealing of course with chromosomes.

We believe that the change from one species into another during the course of evolution is accompanied by a change in chromosome number and structure. Small wonder then that we were interested in the Sumatran rhinoceros which is believed to be the most "primitive" of the five rhino species. In the early days of our ability to assess the chromosomes of mammals—it began in 1960 and has since become a refined discipline—a single female Sumatran rhino was being exhibited at the Copenhagen zoo. How does one get a blood sample or a skin biopsy from a

(Below) Family of Southern white rhinoceros at San Diego's Wild Animal Park in San Pasqual. (R. Garrison, San Diego Zoo)

rhinoceros? We enlisted the help of the late Dr. William T. Schauerte, chairman of the Rhino Specialists Group of the IUCN. He proposed to the Copenhagen Zoo director that I be allowed to take a biopsy for study. On the appointed day in June 1969 I was greeted at the Hamburg airport by Schauerte who told me the deal was off. The Copenhagen director, he told me, had said it was impossible. I was, however, undaunted by Schauerte's pessimism and so off I went to Copenhagen. Yet despite my best efforts at persuasion (enhanced by much whiskey) the zoo director remained skeptical. But the long-time keeper was more accessible, and so in the early hours of the next morning, armed with buckets of apples and loaves of fresh bread, the keeper and I stole into the enclosure. Appeasing the animal with apples, the keeper kept the rhino occupied while I took out my alcohol sponges in order to cleanse the soft skin behind the ears. Then—on a prearranged signal—I reached out with my weapon, and took the biopsy. The rhino, mercifully, took off in one direction while we took off in the other, biopsy in hand. The weapon, concealed until the last minute, was, of all things, an old-fashioned forceps used by gynecologists to take biopsies from the tough tissue of a woman's cervix!

Sumatran rhinos are solitary animals in the wild, rarely traveling with any other rhinos except their young. They are superbly adapted to swampy forests and prefer to frequent the 500 m radius or so of their beloved mudwallow. Krumbiegel makes the unusual observation that in contrast to the behavior of other rhino species, this animal sleeps with a front leg extended under its head. At least the specimen in Copenhagen did. The few scattered populations remaining in Burma, Thailand, Sabah, and West Malaysia each contain only a few dozen or less of Sumatran rhinos and they are widely separated. Add to this the solitary nature of the animal and it will be easy to understand why no reproductive success has recently been possible. This is judged to be the case from the absence of juvenile hoofprints in recent years. Its last retreat is in the northern Leuser reserve (6,000 square kilometers) which contains between 30 and 50 animals and conservationists hope that this at least represents a viable population for future breeding.

In the early eighties 11 zoos came together to map out a last-ditch formal proposal to the Malaysian governments for the capture of 12 pairs of Sumatran rhinos for safekeeping. Surveys commenced that were followed by endless negotiations that eventually ended with an agreement in 1984. Indeed some rhinos were captured by the Malaysian Game Department and one male is now on exhibit in the Malakka Zoo of Malaysia. However, at the time of this writing the government of Sabah (East Malaysia) has not agreed to the exportation of any animal to even the best equipped and most cooperative Western zoos. And indeed in Sabah over one-half of the 30 rhinos existing 5 years ago have been poached, while the destruction of the remaining forests is also an assured prospect. At one time practically all of the Malaysian peninsula was tropical forest. Now only 55% remains and of this, one-third has now been logged. Habitat destruction is, as we have seen, a major cause of most animal endangerment. This is certainly the case for the Sumatran rhino. The two opposing forces in the conservation debate—*in situ* and *ex situ*—must look on while condor and black-footed ferret, and perhaps the Sumatran rhino population, dwindle in the wild to unsustainably small numbers. How much better, we can see, is the prospect for the great Indian and the white rhinoceroses whose enlightened management, translocation, and placement into secure national parks and zoological gardens has given us hope that our children will someday be able to behold these truly spectacular relics of bygone days.

This unusual animal's native habitat is Vietnam and Laos. The douc langur is universally agreed to be the most colorful mammal in existence. Because of its adornments it is called *Kleideraffe* in German or "clothed monkey," as its various colorings are likened to boots, socks, shirts, jackets, etc.

 African colobus monkeys and langurs form a separate subfamily among the Old World monkeys, the Colobinae, a Latin word that describes the near-amputated nature of the thumb of its spectacular black and white African members. More often Colobinae are referred to as "slender" monkeys, but most properly they are grouped together as the "leaf-eating" or "leaf" monkeys because of their staple diet of leaves and buds from a wide variety of trees and shrubs. There are over 30 species in this group, most being distributed over India and Asia. The name "langur" is derived from the Hindi word "langoor" which means "long-jawed." In Bengal lives the sacred Hanuman langur, which ranges high into the mountains, even into the snow, and its footprints may well be those attributed to the abominable snowman.

Douc Langur

(*Pygathrix nemaeus*)

The douc langur is rarely seen in zoos, and keeping any leaf monkey in captivity has been, until recently, a very difficult task as it was not until the past 2 decades that their dietary needs were properly understood. It is a relatively large monkey whose body may be more than 2 feet in length with a tail of equal size. As in all Old World monkeys, the tail is not prehensile. Only some of the larger South American monkeys have such tails which serve them as an additional limb. The reason for the long tail of langurs is unknown. Perhaps it helps in the long (6 m) leaps that they commonly make when traveling through the forest.

Douc langurs are distinguished by their slanted eyes, which give them a distinctly Asiatic look. Of course, this morphologic similarity to Asiatic humans is entirely coincidental, but it is nevertheless striking. Their face is a yellow-brown that fades to white on the sides, and the more the animal is exposed to the sun the darker yellow the facial skin becomes. A subspecies, *Pygathrix nemaeus nigripes*, named after the black feet that both species possess, has a much darker, nearly black face. This subspecies is perhaps already extinct inasmuch as there have been no recent sightings and none are in captivity. Male and female doucs have prominent long white hair on the sides of their faces and a beard that is more pronounced in the males after they reach puberty, at about 5 years of age. The neck and upper chest are white and shade into yellow-brown, giving the impression of a napkin. The major portions of the body are covered by a long, silky haircoat that can be described as "salt and pepper" gray, with black and white hair. Thighs, hands, and feet are jet black. The forelegs are chestnut brown and the forearms white-gray. The long tail is largely white, and a prominent white triangle is seen over the insertion of the tail. In males, this is topped on either side by a white dot. But despite such truly spectacular coloring, their facial expression often seems mournful and aloof. Juveniles are not quite so colorful, except for

(Above) Typical resting post of douc langur at San Diego Zoo.

their facial skin, which remains a light blue for the first 5 months or so of their lives.

Leaf-eating monkeys have been very difficult to keep in captivity and douc langurs, as well as the related proboscis monkey from Borneo, have been among the most delicate and precious charges of zoo curators and their veterinarians. The reason for this, as we have already noted, is that their complex diet is very different from that of most other, generally hardier primates. And it was not until 1964 that this situation was changed dramatically as the result of a remarkable study on the structure and function of the langur stomachs done by the German primatologist Kuhn. He found that, in contrast to the single chamber of most monkeys' stomachs, those of the Colobinae had, with minor variations, three or four chambers. The first two chambers, the presaccus and saccus, are small at birth but soon enlarge as solid food is ingested, beginning at around 2 months of age. The second chamber, the saccus, becomes enormous as it fills with a mush of finely chewed leaf fragments, cut to a tiny size (no more than 3 mm) by the langur's specially modified teeth. In this chamber a large number of mostly unknown bacteria ferment the leaves, much as occurs in the large stomach chambers of such ruminant animals as the cow. But, in contrast to the cow, no ciliated organisms participate in this digestion and the "cud" is not rechewed. Also, unlike our own acidic stomach contents, the contents of the langur's saccus are neutral. Only as the fermented material slowly moves further down into the digestive tract are the bacteria killed by an acid environment that releases gases. This release leads to frequent eructation and burping during digestion and to terribly large abdomens. The fermentation process eventually yields large quantities of fatty acids on which the nutrition of doucs relies heavily, rather than on the sugars and proteins used by other species of monkeys. So they are highly specialized in their digestion and it is probable that social contact with adults is necessary for the young, in order to provide them with the proper bacterial stomach flora without which the digestion of leaves is impossible. We believe that when the young are being breastfed by their mothers they also occasionally receive prechewed, predigested, and egested food from them to prepare their stomachs for the task of leaf digestion.

Death from gastriectasis or "bloat" (as it is well known in cows) was a common occurrence in langurs before adequate studies of their digestive functions were made. When animals were captured they were given bananas, fruit, rice, and other highly carbohydrate concentrated foods. Their stomachs could not handle the gases produced by the fermentation of the rich sugars. In cows, the veterinarian puts a needle into the distended stomach, lets out the gas, and often the animal re-

(Below) Douc langur mother with month-old male infant.

covers. Such treatment is too drastic to be done on frail monkeys; instead we have simply had to learn exactly what to feed them. At the Basel Zoo, for instance, which has kept douc langurs since 1969, over 40 species of leaves and bushes comprise their diet. In the Cologne Zoo, where a healthy group of douc langurs has existed since 1968, the eminent primate curator, Uta Hick, has carefully evolved diets for all of these species and their young, which most zoos now follow. For while animals in the wild clearly select not only what is most palatable to them, but also what is "safe," such discrimination may not be possible in captivity. Thus, while acacia leaves are readily consumed, not all acacia species may be safe. We know that leaf stems and other stringy cellulose material pass undigested through the langur stomach. And we have had the bitter experience of witnessing some intestinal perforations in Indian Hanuman langurs. A clever botanist who examined the cause of perforation identified spiny parts of young acacia leaves as the culprit, the same material that sometimes causes obstruction by matting together and forming bezoars (concreted mass found in the stomach of certain animals). So at the moment, only nine zoos around the world have a total of 50–55 douc langurs, and it seems highly unlikely that new animals will be coming from the wild to enlarge these numbers. Vietnam stopped exporting all primates in 1970 and they are now a "protected" species there.

But what is their status in the wild? The new Vietnamese government has expressed concern over its dwindling wildlife and a few parks are in the planning stage. Yet the real status of langurs in the wild is truly unknown, except to say that one member of the subspecies, the blackfaced *Pygathrix nemaeus nigripes*, is severely threatened or perhaps already lost. And although the douc is "protected" in Vietnam, concerned officials in that country know that the animal is frequently hunted as a food source. It is said to be eaten only by men, women considering monkey flesh unclean. This practice is particularly surprising when in West Africa, for instance, the colobus monkey is revered as messenger of God, a reputation enhanced by its habit of sitting in treetops at sunset. The Indian Hanuman langur is considered so sacred that it is allowed to roam undisturbed in large bands through temples and towns. Likewise, the spectacular Chinese golden monkey is revered in that country. This monkey was first described to us by Père David, the discoverer of Père David's deer. So it is therefore difficult to understand why the closely related and quite spectacular douc langur should not be equally esteemed.

However, more damage may have been done to the douc during the war with bombing and particularly with the defoliation of as much as one-sixth of its habitat than by anything else. Although the damage to these jungles has gradually been repaired, the reproductive capabilities of the douc are not great enough for a quick repopulation of the area. So here is a true challenge for zoos: to bring the numbers of captive animals up to levels that will guarantee "captive self-sustaining populations," the new watch word for modern zoological gardens!

Bats truly need all the help they can get. They suffer more from bad press than from just about anything else. In general (in the West) people fear bats. This is not withstanding the many interesting folklores about bats, from American Indian stories to Aesop's fables. In Oriental culture, however, they are viewed as a symbol of happiness. Whatever the reason for this dichotomy, the vicious pursuit of bats in the West has led to a drastic reduction in their population. So critical is the situation that, in order to educate the public to a more benevolent attitude towards bats, Dr. Stebbings of England and Dr. Tuttle of the USA felt it necessary to form an organization called "Bat Conservation International" in 1982.

Although we don't see bats very often because they are nocturnal animals, it has been estimated that as many as 10 billion of them take to the skies each night in order to perform (although this is little known) various vital tasks in the ecology of our planet. Some feed exclusively on insects, which, if left unchecked, might overwhelm us. Others pollinate plants and spread tree seeds over areas that have been deforested such as the mangrove swamps.

Bats

(*Chiroptera*)

Although there are about 1,000 different species of bats, they all belong to one order, Chiroptera. This word has a Greek root and stands for "hand-winged." Chiroptera are the only "true" flying mammals. There are many widely differing phenotypes of bats, meaning they have widely differing external characteristics and equally diverse habits. There are fruit-eating, insectivorous, flower-feeding, canivorous, fish-eating, and, of course, vampire bats. They also differ widely in size, from the 2 gram bumble bee bat of Thailand—one of the 12 most endangered animals in the world—to the 900 gram (2 pounds) fruit bat with a 1.5 meter (5 foot) wingspan. Such diversity has led to the separation of bats into the suborders Megachiroptera (big bats) and Microchiroptera (all the rest). But it is perhaps easiest to identify them by their individual family names, as these describe some of their overriding behavioral or anatomical characteristics. And so we can speak of guano bats, flower-nosed, spear-nosed, hog-nosed, long-nosed (flower feeding), fishing, thumbless, butterfly (white spots), hammer-headed, big-eared, epauletted, free-tailed, sheeth-tailed, and tent-building bats to name just a few. For a much more detailed survey of the various types of bats I strongly recommend Nina Leen's beautiful book *The World of Bats* to the reader along with a free subscription to *Bat News* which is published by the Fauna and Flora Preservation Society of the London Zoo. Your efforts will be well repaid with much fascinating information.

Bats are haired mammals, with even the outer surface of their transparent wings bearing fine hair. While many of them are brown, gray, or some other dark color, there are yellow bats and white bats (known as ghost bats). Others have white spots (the butterfly bat) while some (the male epauletted bats) have complex yellow shoulder pads. The most widely known characteristic of bats, though, is of course, their wing membrane. The much elongated bony members of their hands and feet are connected by this membrane, which is very thin and extends essentially from thumb to little toe. This thin membrane carries within it three major blood vessels and, being so delicate, when injured it heals rather poorly. Their bones are also very thin in order to allow flight, and a fracture often means rapid death. The wing membrane is folded when the bat suspends itself, using its hind limb claws, in caves, trees, or rocky ledges. It protects the animal from cold, heat, and rain. Most have very large ears, which act as the receivers for their echolocation sounds and their noses and mouths are all well adapted to the various tasks that each species performs—nectar collecting, echo emitting, or fruit biting. Canine teeth are often prominent, as are the proverbial "chisel-like" teeth of the vampire bat.

Many tropical bats reproduce throughout the year. But in temperate climates the adults hibernate and copulation usually takes place in the fall. The sperm is stored in the females over the winter, and fertilization of the ovum takes place after hibernation. It is supposed that the males are too weak for copulation after their energy resources have been almost consumed in the hibernation period, which can last up to 6 months. During this hibernation period bats are quite helpless and subject to mass destruction. If they are disturbed during this time the energy consumed in "waking up" and reentering the hibernation state may so reduce their barely sufficient fat stores that they will not last to the winter's end and up to thousands may perish. Merlin Tuttle, bat aficionado and mammal curator of Milwaukee's Public Museum, relates impressive though tragic instances of rapidly declining bat populations due to such disturbances. He describes the disappearance of a colony of over 250,000 in Alabama whose cave could have been reached only by boat. In the past this colony had set out in a swarm 60 x 30 feet that would last for hours while the bats hunted insects. A colony such as this may devour as much as a million pounds of insects per night! So it is easy to understand why the demise of these "natural insecticides," as they have sometimes been called, might have disastrous long-term effects on our ecology. It has been estimated that the Texas guano bat consumes 6,000 tons of insects annually!

And at one time the guano produced by these bats was "mined" and sold as fertilizer. During the Civil War it even served as a starting material for gun powder. But the loss of insectivorous bats would be severely felt by civilization, as they serve to control many of what we regard as insect pests. This was at one time an appreciated fact because in times previous to our own, entrepreneurs sold "bat towers" to American farmers in order to help them control insects that were harmful to their crops.

But our present ignorance of the true nature of bats extends into some other areas. For instance, everyone is familiar with the saying "blind as a bat." Yet nothing is farther from the truth. For all bats have eyes and can see. Those fruit-eating bats that are without the echolocation system possessed by most bats actually have large, protruding eyes and can see quite well. These bats often travel as far as 15 km to the orchards and they locate their food by using their sight and sense of smell, both of which are well developed for this task.

The ultrasound used by most bats for echolocating insects and obstacles is made by the larynx and is transmitted to the "sender." In most species the mouth serves as this "sender" but, for instance in leaf-nosed bats, the complex nose appendage is used. Bats can eat while echolocating which is quite a clever development. The ultrasound impulses (which generally cannot be heard by the human ear) hit the insect or obstacle in rapid intervals, are reflected back, received in the bat's unusually large ears, modulated, and then processed in the brain. Intensive research has come a long way in being able to characterize this complex and variable system. The smallest insects, the direction of their flight and their speed, can be accurately analyzed by this auditory system as if it possessed a computer. The fact that bats are rumored to become entangled in women's long hair is entirely incorrect. They are only trying to catch the mosquitoes that encircle our heads at night, so be grateful! The echolocation systems of the vampire bat and the carnivorous bats, however, function somewhat differently. Their ultrasound has a much lower ($1/1000$) intensity and small movements in their prey can be detected. Thus a vampire bat, starting its foray well after nightfall, flies a few feet above the ground hoping to find any large mammal—a cow, horse, pig—and when it detects one, the bat comes gently to rest next to it. At this point, vision takes over and the bat hops up to the resting beast and finds an exposed area for bleeding. The three vampire bat species of South America, really quite small animals, are so skilled at their task that their prey do not even feel the skin incision of the bat's razor-sharp teeth. Once a vessel is opened, they lick up the blood, their saliva having an anticoagulant capacity to keep it flowing. Their tongue is rolled into a tube that they use to suck the blood until they are gorged and unable to fly away. They digest near the donor animal until they are ready for the flight home.

But I digress. I was speaking of echolocation. It has long been known that bats can avoid obstacles as thin as a human hair, even when blinded. But only recently was it learned that they perceive moving objects significantly better than stationary ones. Even perhaps more astonishing is the fact that some moths, as a defense, have developed ears with which they can hear the echolocating signals of bats—and these moths escape being eaten more often than those species that have no such ears. Indeed when these moths hear the strong signals of incoming bats, they may dive and lie still on the ground, much to the chagrin of the hungry bat. Still other moths have adapted to make ultrasonic clicks which, so we suspect at present, may "jam" the processing of echos in the bat's brain. Or it may "misinform" the bat that it is really an inedible critter. There are many such adaptations and scientists, armed with ever more sophisticated equipment, have a great deal of fun investigating the confusing array of bat–insect interaction.

But why then, if bats are such efficient hunters and insects are so abundant, do they need protection? Well, the problems bats face differ, depending upon what type of bat you choose. Fruit bats, for instance, because of their rather large size, are hunted for food in the tropical countries they inhabit. During their sleep—a stage of torpor with reduced metabolism—they are easily caught or shot. They are also persecuted under the mistaken belief that they raid fruit plantations. While they do eat bananas or mangoes from plantations when these fruits are ripe, most commercially valuable fruit is picked when it is still unripe and the bats will not eat unripe fruit. Moreover, by and large, they prefer wild fruit anyway, but perhaps the main problem for Megachiroptera is the wholesale deforestation of their resting places. And, to make matters worse, fruit-eating bats are essential in the dispersal of seeds for regrowth of the jungle. In Borneo and other places their guano has been shown to be vital for the maintenance of mangrove swamps. The flying foxes of Queensland have been persecuted because they were believed to be fruit eaters, but they are in fact primarily blossom feeders and, like nectar drinking bats, they serve an essential function in plant pollination. For the billions of insectivorous bats the problems again are primarily interference with their habitat. As I mentioned earlier, hundreds of thousands of bats in caves can be destroyed within hours by man's intrusion. Quarry mining (as in Thailand), cementing of caves, and careless entering (spelunking), as well as the willful destruction of colonies, all play a major role in harming insectivorous bats. Wood preservatives sprayed in attics have been found to be significant hazards for bats in Europe, where some authorities fear bat extinction will occur within the next decade. And while this may be overly pessimistic, no doubt exists that there has already been a drastic decline of European bats in recent years. Consequently all bats are now protected in Germany. And surveys conducted by Bat Conservation International show that, once accurately informed about bats, 75% of the people so educated express a change in attitude and a loss of fear toward them. I trust the women reading these pages will once and for all accept the fact that bats will not entangle in their hair.

But what about rabies? Of 226 cases of rabies diagnosed in 1983 only 8 (or 4%) could be possibly attributed to bats. House pets are a much more likely source for this disease. Inevitably we must see that bats are beneficial to us. Just imagine the profusion of insects we would have if we lost them! Many bats in fact have been kept and trained as pets, as they are really quite gentle animals.

Of course we can do nothing personally for the bumblebee bat of Thailand, one of the 12 most endangered animals in the world today. But one wishes that this tiny mammal, discovered only in 1974, could be embraced by conservationists. It is probably too late for this species. But that doesn't mean we should be oblivious to our own bats. Read *Bat News* which I heartily recommend to you. It will give you a wonderful insight into the fantastic night-flying world of these creatures. And long live the bats!

When we think of vanishing "animals" we generally tend to forget about all those species that are not mammals. But butterflies are also animals, and they may be just as endangered as the famous small darter fish once was or as the panda now is.

Butterflies and moths form one order of insects, the Lepidoptera. There are at least 15,000 species of butterflies and many more of moths. In the United States alone we have some 7,500 species comprising seven families of butterflies and 75 families of moths. Butterflies are generally characterized by having two pairs of wings, a proboscis for drinking nectar which can be rolled up, and specialized antennae for the reception of pheromones—volatile hydrocarbons used for species-specific attraction. By and large butterflies have antennae with bulbous ends while moths have branched and highly complex antennae. Lepidoptera lay eggs (moths in rather larger numbers than butterflies) from which their larvae (caterpillars) develop. These caterpillars then feed on highly specific food plants and so consequently can at times be considered pests. But the highly specific nature of their needs restricts their distribution and at the same time makes them vulnerable to extinction if the food source should be destroyed. These requirements also make keeping them in captivity a difficult task.

Butterflies

(*Lepidoptera*)

Butterflies owe their often very vivid colors to the reflection of light on the innumerable tiny scales that cover their bodies and to a variety of pigments that cover these scales. The scales develop as the caterpillar is transfigured within its "chrysalis," or cocoon, into the adult butterfly. Cocoons have just as many highly characteristic shapes as do the often fierce-looking caterpillars, each particular to its species. These scales also give butterflies and moths the name Lepidoptera, as it is Greek for "scale winged."

Selective breeding of butterflies now occurs in many parts of the world. All over Europe, and particularly in England, butterfly houses are being created to enhance breeding and study. Many can be visited and the emergence of a butterfly from its chrysalis, the unfolding of its wings, their drying and taking off are now events that can be witnessed by appreciative audiences. No doubt this is a powerful stimulus for conservation education. Likewise the insectarium of the Cincinnati Zoo provides unexpected pleasure to thousands of visitors each year. But "ranching" or captive breeding of butterflies is not, as we have already stated, easy to accomplish. Many different types of food plants must be maintained for the many different species. They need to be grown in large quantities and kept free from insecticides (ants love to prey on the chrysalis) and they must also be kept free of herbicides and heavy metals. Visitors to such institutions are often encouraged to buy seeds for the many grasses and weeds needed by different species of caterpillars. Purchasing the seed has encouraged the reestablishment of food plants and, remarkably, many butterflies can become more abundant. The butterflies' excellent sense of smell allows them to seek out such plants and, because of pheromones, this keen sense also allows them to seek out a mate.

(Left) The microscopic appearance of the wing of a common white cabbage butterfly with its darker brown patch near the branches of the veins. The pattern of color and veins is highly specific.

(Above) The edge of a wing with fine hair, also showing the scales.

Pheromones are interesting substances. The name "pheromone" was coined only in 1959 when the German chemist A. Butenandt first discovered it in moths and named it "bombykol." He had to extract the abdomens of some 500,000 moths to produce 6 milligrams of this chemical. It is produced and stored in a specialized gland in the female. To attract males, the female empties the gland by elevating her blood pressure, thus discharging the contents. Certainly over many hundred meters the males can smell the substance and assume a directed flight toward the source. Lepidopteran pheromones have since been studied by much more sensitive techniques and far fewer insects are needed in order to isolate the pheromone. At the Institute of Organic Chemistry in Erlangen, Germany, scientists, headed by Professor Bestmann, were able to put microelectrodes onto the antennae of moths, and, using highly complex amplification systems, they recorded the signals received by male moths when they were exposed to small numbers of pheromone molecules sent through a wind tunnel. It turns out that the pheromone molecules that excite males are short olefins, hydrocarbon chains that are received in the hairs of the male moths' antennae via tiny tubes that end on a very sensitive receptor. They are then transduced to the neural system and there the signal is processed. As little as one molecule can be received, but for the appropriate reaction the male must receive a number of them. Once their structure is known, these pheromones can be manufactured synthetically and with such artificial compounds one can attract male moths of a given species. They can be lured into sticky traps and destroyed, providing they do not sense an artificially high concentration of the pheromone. Related compounds can also be produced that will simply confuse the male moths so drastically that they will never be able to find a "real" female to fertilize! There is now, as a result of such research, real hope that we will be able to very selectively control damaging species or, for that matter, enhance the breeding of endangered ones. For all these reasons the creation of a "Captive Breeding Institute" for butterflies has been called for, in which all such aspects could be studied and in which conservation efforts and the study of Lepidoptera could flourish.

(Left) Northern hairstreak, *Strymon ontario*, a relatively rare butterfly occurring from Canada to Texas. It is a typical "hairstreak" because of the fine hair at its end and possesses the club-shaped antennae of butterflies. Its early stages are poorly known, but the larvae prefer hawthorn leaves.

(Above) The Northern American luna moth, *Actias luna*, is very pale green and the people's choice for the most beautiful moth of the region, the eastern half of the United States. Males appear similar to females. The green caterpillars prefer walnut, beech, birch, willow, oak, and hickory. The huge, typically moth-shaped antennae can smell females from miles away.

And there is another fascinating aspect of butterfly biology that I would like to touch on, and that is their migration, an event well documented in the North American Monarch. While the average life span of a butterfly may be only a month, some butterflies, like the monarch, have time to migrate. They winter in temperate climates and then make a partial return trip to the place of their birth. The monarchs of the northern states and Canada make the long journey to specific remote mountain forests in Central Mexico and California for they winter in large swarms. Perhaps as many as 50 million or more roost in thick masses in fir trees in semihibernation. In spring they will venture north once again but few will reach their point of origin. The females lay eggs and the hatchlings will continue the journey by some mysterious instinct, only to return to the wintering grounds in the following fall. Once they have been identified these highly specific wintering forests obviously need careful protection lest one of our best-known butterflies disappears. Such protection is the business of the "Pro Monarca" branch of the Xerxes Society. Many other butterflies migrate, and some even associate in mixed swarms. The Painted Lady of Europe is a prime example. It crosses the Mediterranean to and from Africa in large swarms and is sometimes seen out at sea. Its American relative, also called the Painted Lady, shows no such inclination. Occasionally however, some of these migrations go astray and that is perhaps the reason why our monarch may sometimes be found as far away as Australia.

(Above) Underside of Northern hairstreak. Typical for most butterflies are the two pairs of wings. Note also the zebra-striped antennae stem.

And it should be pointed out that not all caterpillars should be considered pests as is often the case. In fact, the extensive pollination of plants undertaken by these insects during their hunt for nectar all but makes up for the harm they do to some food plants. And science has learned much from watching their evolution in progress. Rarely has this been possible in other animals. Thus, the European peppered moth was once white with some dark spots. But, as industrial pollution made houses and trees darker, the light-colored moths became easy targets for predation in their daytime resting places. Gradually, however, a darker form, at one time extremely rare, has now nearly completely replaced the former white moths, obviously an adaptation to the changing environment.

But this small chapter has been only a very fleeting introduction to what are some of the most beautiful animals of our world. We can heartily recommend that you pursue an interest in these "jewels of the sky" for they will provide you with endless pleasure if given enough attention. And try to remember the next time you are mowing your lawn, to leave a few weeds around and in that way you can make your own contribution to the conservation effort!

"It has been justly remarked" said C.P. Groves, "that the taxonomy of the genus *Gazella* is one of the most confused in the whole class Mammalia." In the last century the horned ungulates were divided into the various bovids then recognized: the sheep, goats, oxen, and all the rest, the antelopes. It soon became apparent, however, that the latter group included so many divergent animal forms that subdivision was necessary, and even today scientists are not finished with this task. Periodically they regroup different species of antelopes because new features have been recognized, skull shapes have been interpreted differently, or for other apparently cogent reasons that escape most of us with only a peripheral interest. Gazelles are, however, true antelopes and modern zoologists such as Walther recognize some 17 species. But whatever questions remain concerning the descendancy of gazelles may remain forever unanswered since a number of species have already become extinct and others are near that point. Most studies on gazelles can presently be conducted from skins and skulls only and often their origin is imprecisely or totally unknown.

Sömmerring's Gazelle

(*Gazella soemmerringi*)

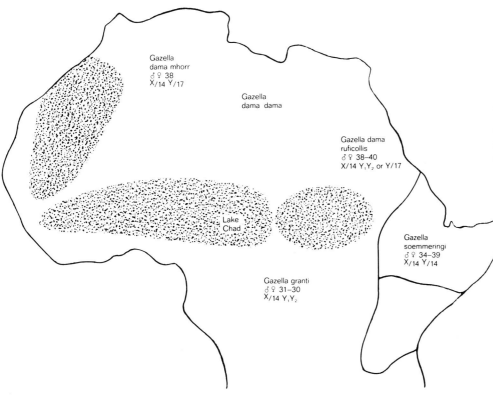

This gazelle was first described by Cretzschmar in 1826 when he was describing Rüppel's travels through North Africa, and the name Sömmerring was attached to it in honor of that scientist's 50th doctoral anniversary. It is the smallest of the three species of the "Nanger" group and is confined to the horn of Africa, essentially Somalia. Normally they live in groups of five to twenty but during the rainy season groups of up to several hundred gazelles may trek together into the desert in search of new food sources. Ordinarily they live in the hilly savannah, punctuated by Acacia trees. They eat grasses, herbs, and buds and pair in the fall to have one, and rarely two offspring in April. But they are so elusive and it is so difficult to gain access to their habitat that few biological studies of them have been undertaken. What we do know of them comes largely from the few animals that have been kept in zoos, the best study of their behavior being done in the Hannover Zoo in Germany. From 1960 to 1970 this zoo produced 14 offspring. But this group, as we shall see in some detail, died out. A similar fate also befell a second attempt in Hannover during the present decade.

In its native Somalia, the Sömmerring's gazelle has recently diminished remarkably in numbers because of competition with the grazing domestic livestock of an expanding human population, because of hunting, and because of warfare. Gazelles range from eastern Asia (Mongolian and Tibetan gazelles) to Africa, with forms in India and Arabia (the mountain gazelle—*Gazella gazella*). Well adapted to arid and steppe environments, the gazelles of Africa developed most abundantly. But when drought recently threatened millions of Ethiopians in the infamous Sahel region of North Africa, it is easy to understand why fragile gazelles, however well they may be adapted to a waterless existence, should also suffer. The Nanger group, those animals with white backs and prominent scent glands beneath their eyes, are perhaps the most severely threatened of all. The most widely distributed species of the Nanger group is the Dama gazelle, which, so we believe, does not reach east of the white Nile. In Morocco its appearance is so dark that it has long been regarded as a "good subspecies." It is also known there as the "Mhorr" gazelle, a name derived from the Arabic word *mhur* which connotes a young animal. But it is extinct in the wild now. The last wild one was probably shot in 1968 and, were it not for the efforts of the Spanish Major Estalayo who enjoyed keeping wild animals in his yard, we would know little about this gazelle. In 1971 he shipped nine animals to Professor Valverde in Spain, who in turn placed them in the capable hands of Dr. Cano at his Almeria compound. From these and other breeding animals acquired in 1975 a total of 67 gazelles had been produced by 1980 in Almeria, thanks to the efforts of Dr. Cano and his daughter. It was then necessary to disperse some animals, lest an accident, such as an infection, pose a serious threat to this small breeding stock. Thus in 1980 various zoos received breeding stock; San Diego has had 12 offspring from the four animals they received but only seven have survived. Is inbreeding the cause of the heavy mortality? What other factors may be at work?

The Dama gazelle, whose origin is the vicinity of Lake Chad, also belongs to the rarest specimens seen in zoos, and since 1983 it has "achieved" the status of "threatened" in the wild. It has disappeared from many areas where formerly it was seen in abundance and some authorities believe that its nominate form, *Gazella dama dama*, is locally extinct. The most widely kept form of Dama gazelle is the red-necked, or eastern Addra gazelle (*G. dama ruficollis*). It usually lives in troups of 5–10 animals, except in the rainy season, when hundreds congregate to march hundreds of kilometers in search of new vegetation in the desert. Perhaps it is during such a migration that the differently colored subspecies mixed. Also perhaps the reported occurrence of these different forms from the same areas is due to the fact that they were first described at a time when migrations had dislocated animals from their usual habitats. We will never know. Fortunately, however, in 1967 23 red-necked Dama gazelles were brought into zoological gardens, principally into the efficient breeding facilities of the Catskill Game Farm in New York and to the San Antonio Zoo in Texas. By 1981 there were 178 in captivity and many more have been bred since then. Although capture and transport are difficult—the animals can reach speeds of up to 80 km/h–once adapted to adequate zoo facilities they do well.

But with Sömmerring's gazelle we may not be quite so lucky, perhaps because of their rather complex genetic composition. Ten years ago we studied the chromosomes of the 13 species of *Antilopinae* then available to us, and we noted them to have great differences. The most remarkable aspect was that nine species had chromosome numbers that differed from male to female, the males always having one more element (e.g., Dorcas gazelle: male 33, female 32). In only four species (impala, gerenuk, springbok, and Thomson's gazelle) was this not the case. It turned out that the uneven number of the former nine species was caused by the attachment of one sex (X) chromosome to a non-sex (A) autosome. Because this is a most extraordinary event in mammalian genetics one must postulate that these nine forms all had one ancestor in their

distant past who first "invented" the aberration and then passed it on to all descendants. Interesting though this was, we absolutely had to have the one species still missing from our list, the Sömmerring's gazelle. We have since studied 28 of them with the most unexpected results. Yes, they did have the X/A fusion, but they also had chromosome numbers varying between 34 and 39. Perhaps this is the reason why their reproductive success has been so poor. Already, as we have seen, the two groups in the Hannover Zoo died out. Our San Diego animals came from Busch Gardens in Tampa, Florida where seven animals had been imported between 1966 and 1970 from dealers in Germany. Are they all Sömmerring's gazelles? Or are they hybrids? In the regions of northern Kenya where the Sömmerring's and Grant's gazelles meet these two species are said to resemble each other so closely that even experts cannot tell them apart. Do they perhaps also meet some Dama races in their long migrations and reproduce with them? We truly do not know.

To find an answer to what real Sömmerring's gazelles are we may have to look at the animals that have been isolated on an island off the Somali shores where a dwarf Sömmerring's gazelle exists.

Six of the smaller gazelles were in very serious danger of extinction in the wild earlier in this decade and their demise is considerably more probable now due to the prolonged drought in the Sahel, the rapid population increase, and the continued warfare in the area. The Sömmerring's gazelle was chosen here as a sentinel species, a vanishing species whose biology is poorly understood and for which little time remains in which to rectify its dismal situation. How important an understanding of the basic biology of animals is for their captive management should be readily apparent from these pages. But although one may hope for an improvement in the local habitat of these lovely animals, there is truly no room for uninformed optimism!

(Above) Thriving group of red-necked Dama gazelles (Gazella dama ruficollis), also called Addra gazelles, at San Diego's Wild Animal Park. (R. Garrison, San Diego Zoo)

The order Chelonia (meaning double shelled) or *Testudines* (Latin for tortoise) is ancient indeed. Its members had essentially similar forms 150 million years ago. They are widely distributed and fall into two main groups. The members of one suborder, the Cryptodira, have the ability to retract their necks into the shell. There are 11 families in this suborder, with 180 species that are found mostly in the northern hemisphere. The second suborder, *Pleurodira*, cannot retract their necks and are found only in the southern hemisphere. The Galapagos tortoise belongs to the former group and is one of the giant tortoises that may weigh up to 600 pounds, grow to 5 feet in length, and that lives probably well over 150 years. Most taxonomists use the word tortoise to describe those *testudinae* that are exclusively terrestrial (*Geochelone*), the term turtle being a more encompassing designation. The word tortoise also has its roots in another Latin word, *tortus*, which makes reference to the animals' twisted legs, and with the designation *elephantopus*, reference is made to the enormous dimension of their feet from which only tiny nails protrude.

Galapagos Tortoise

(*Geochelone elephantopus*)

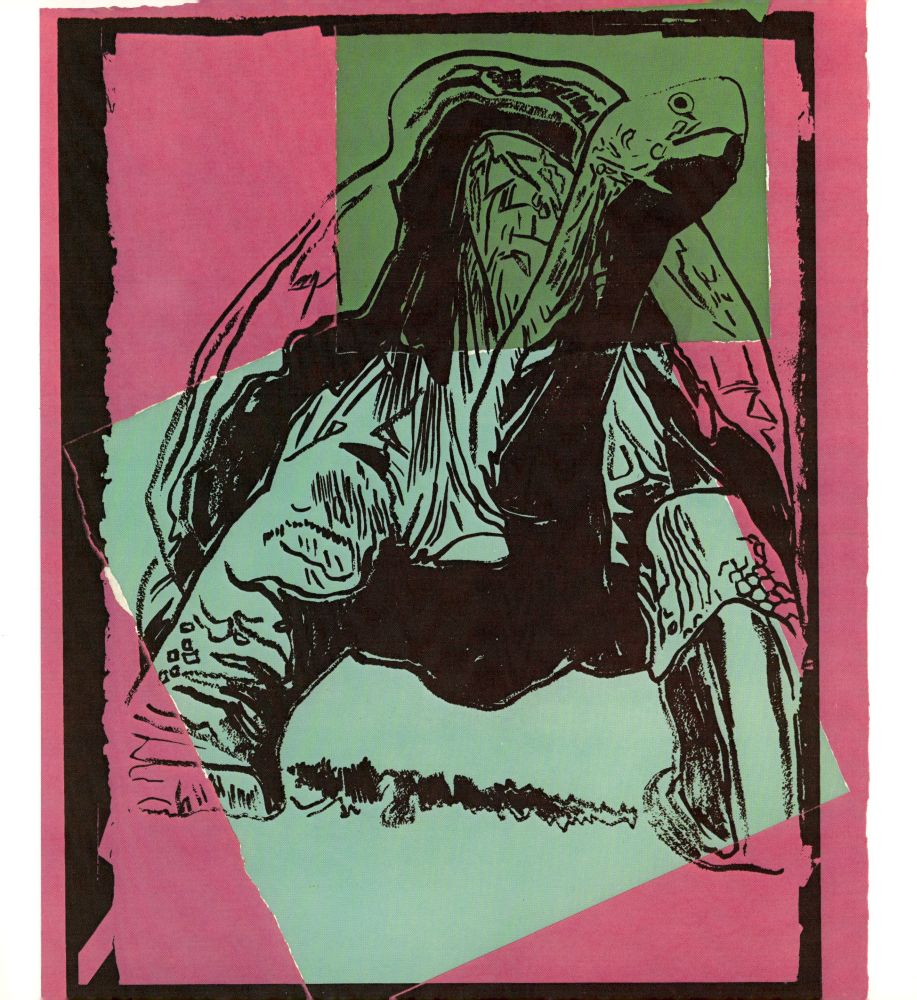

Galapagos tortoises come from the 13 islands of the Galapagos Archipelago, which lies some 1000 km off the shores of Ecuador in the Pacific Ocean. While the islands were first discovered by the Spanish bishop Tomas de Berlanga, they became truly renowned only after the 5-week visit of English naturalist Charles Darwin in 1835. It was as a result of this visit and his study of the isolated fauna of the various islands that Darwin's concept of evolution developed. The governor of the islands was the first to direct Darwin's attention to the different shell shapes of tortoises from different islands. Darwin ultimately came to view these differences as adaptations by the animals to the various living conditions prevailing on the island to which each was specific. For instance, the larger dome-shaped animals came from the larger islands with heavier vegetation. The tortoises with flatter carapaces, those with a "saddle" back, came from islands with less abundant food. They had apparently evolved the ability to curve the anterior portions of their carapace upward in order to create a larger opening for their necks. Thus, their necks can be extended farther to permit feeding on plants that are higher, such as cacti. The ritualistic battles for heights, with necks extended way into the air, are spectacular events in the lives of these animals. And this extension of extremities and neck can also be "called for" by a specific behavior of some of the notorious Darwin finches. Such extension allows these birds to clean the tortoises' skin of ticks, thereby achieving a mutually beneficial symbiosis.

When the tortoises were first discovered, they were abundant on all the islands and their numbers were perhaps limited only by the availability of food resources. They had virtually no predators. But sailors found the tortoise meat to be delicious and the fat to be most suitable for cooking purposes. In the days before refrigeration and

(Above) Young Galapagos tortoise hatching from its shell at the San Diego Zoo.

canned food it was found practical to take tortoises aboard ship for long-term provision. They would live for over a year without needing food or water and they would still, after that length of time, provide a delicious meal. The logs of whaling ships indicate that up to 500 at a time were taken aboard at the regular stopovers these ships made on their long voyages. But the real disaster came when mammals were introduced to the islands. Only an occasional hawk had predated the tortoises' eggs before their introduction. But now the "new" rats and dogs frequently devoured the clutches of 3–20 eggs. Also the goats, pigs, and donkeys brought ashore by the sailors would trample them. But more importantly still, for the first time there was competition for the vegetation on the islands. Thus today only about 13,000 or so tortoises remain. However, on the brighter side, while some 5,000 people presently live on the islands, their behavior toward the flora and fauna there has changed dramatically, and while tourism is an important industry, it is the scientific value of these islands that has now become their most important aspect.

But it has turned out that the mixed herds of tortoises held in the Darwin Research Station and in various zoos have not done too well. The differences in carapace shapes have not only occasionally prevented successful copulation, but have also led to decreased hatchability of the eggs and to anomalies in the young. Once this important factor was perceived, however, and the respective subspecies were separated, hatching success increased.

Tortoise young hatch in about three to four months and thereafter the 2–3 ounce youngster is on his own. That is, he is on his own in the wild; in zoos we tend to hatch the eggs in incubators and feed the newborn. In the wild the young are no match for rats and dogs and therefore conservationists on the islands now carefully surround nests with stones to guard the new crop. In a wet year the female may have up to four clutches with as many as 20 eggs in each. In dry years she may suspend

(Above) Galapagos tortoises at the San Diego Zoo. Note their strikingly differing shell shapes. Presumably these come from different islands of Galapagos.

egg-laying altogether. Once the infants are about 2 years old they are fairly invulnerable, except to accidental falls which may turn them onto their backs and leave them helpless.

Captive management would then seem rather easy, or so one would suppose. You collect the eggs, incubate them, feed the young some cactus, and release them when they are 2 years old. But consider this: how do you get the adults to mate, how do you induce egg laying, and indeed, more importantly, how do you tell the gender of a tortoise? Anyone who is interested in tortoises will tell you that not many external characteristics distinguish males from females. In fact it is not unusual to find a tortoise that has been thought to be a "male" laying eggs! A major problem then, in the propagation of these animals, as well as in other reptiles, is the ability (or inability) to determine their sex accurately. In mammals and birds the sex is determined by the difference of sex chromosomes, XY in a man and XX in a woman. But there are no sex chromosomes in most reptiles, and for a long time scientists did not understand how a roughly equal ratio of males/females was achieved by these animals. It eventually became apparent that the temperature of egg incubation was the deciding factor. When, for instance, sea turtle eggs were incubated at 30°C, half became males and half became females. However, when the temperature was raised to 32°C, all offspring were females, while when the temperature was lowered to 28°C, all became males. Imagine how sad our successors would be if, despite all our best intentions, we were to produce only males. What is worse, we would not even be able to tell what we had produced for perhaps 50 years, when the tortoises reach puberty. And by then it would be too late to begin a "self-sustaining population." So scientists in the Galapagos islands have carefully measured the temperature of natural nests and found them to vary between 24 and 32°C. It seems probable that the nearly equal distribution of sexes is due to female haphazardly laying her eggs, some in sunny places, some in the shade. It also turns out that there are some rather critical hours in the incubation during which the sex is determined, and that this critical phase varies from species to species. Unfortunately, for the Calapagos tortoise, we do not yet have the precise timing or temperature of this event.

However, perhaps the most extraordinary feature of these tortoises is their longevity. To be sure, their hearts beat slowly and they do not have a great metabolism. But then it is quite surprising that their weight doubles annually during their early years. Perhaps their longevity has something to do with the fact that they have no teeth. Tortoises are vegetarians and they cut their food with the very sharp horny edges of their jaws. Anyone who has ever been bitten by a turtle—the snapping turtle comes to mind—knows how powerful their jaws are. This horny ridge perpetually renews itself and never wears out, quite unlike the teeth of mammals that wear down and are prone to decay and injury which often spells death from starvation or infection.

The horny plates so admired as "tortoiseshell" for combs and other utensils or ornaments also renew from their bony underpinning. The "scutes" of the back, so characteristic in the variable appearance of tortoises, and the plastron of the underside have concentric rings from growth. While these are not annual, like the rings of a tree, they do denote growth and the age of the animal. And without these constantly renewable plastrons, the dragging of the tortoise's body as it walks over lava and rocks would soon bare its bones. Visitors to the islands are usually impressed by the well-worn paths that the tortoises have used for eons. Here they move at the rapid speed of 300 yards per hour to and from the feeding grounds and the warm lava pools where they love to lie.

The main threat to many turtles today comes foremost from the collecting of their eggs. As the female abandons the nest immediately after laying, often leaving very obvious trails behind, the eggs are harvested in great numbers for human consumption and also by some natural predators. Fires at nest sites, deliberate gassing, and dune buggies all take their toll, and neonatal mortality is high for all tortoises as we have already seen. Then, of course, there is the famous turtle soup. Their excellence for culinary purposes will no doubt spell extinction for some species. Fortunately CITES now prevents much of the trade in tortoiseshell or stuffed tortoises for ornaments, but although the public is gradually being educated to abhor the acquisition of some items we still have a long way to go. Turtles are not, for instance, the cuddly anthropomorphic animals that many people think they want to have as pets. It is amazing to learn that an estimated 300,000 turtles are imported annually to England alone, and that of these only 5% survive the first winter. Thus, turtles as a group are having real problems, and it should come as no surprise then either that the large Angonoka, the Madagascar tortoise, is now one of the 12 most endangered animals in the world as selected by the IUCN. No wild nests of this species have ever been found and captive breeding has been nearly a total failure. The Galapagos tortoise is a little better off, no doubt because of the great public and scientific importance attached to Darwin's discovery. We can only hope that much will be learned from its intensive study and captive management that will benefit the rest of its endangered relatives.

(Below) The construction of a turtle's carapace. Beneath the leathery scutes are bony plates, tightly interwoven, upon which the horny layer is grown by addition of rings in the periphery.

Of the 8,500 or so bird species in the world the parrots have endeared themselves to man the longest. Not only have they existed for tens of millions of years, but they have also been kept as pets for as long as seafarers began returning from foreign lands. It is even said that a swarm of parrots had something to do with the discovery of America, as it made Columbus change his course.

While parrots are now mostly tropical birds, occurring largely in Australia, New Zealand, and South America (with lesser numbers in Africa and Madagascar) they were at one time found in Europe and the United States. Not so long ago the Carolina parakeet was widely distributed in North America's Eastern states, but, like the passenger pigeon, it has been extinct since the early part of this century. Many other parrot species are not extinct and all of them died because of man's destruction of their habitats and, not infrequently because they were eaten. The Romans served parrot heads as delicacies and fed them to their lions as well. But many parrots are also caught for the pet trade, and many illegally traded endangered species have died as a result.

Puerto Rican Parrot

(*Amazona vittata*)

There are 376 species of parrots, distinguished principally by their hooked beak. They form the order Psittaciformes and are grouped into seven subfamilies. All parrots are characterized by an upper beak that is mobile against the skull and that has a sliding lower jaw. This arrangement allows for fine manipulation of food items. It gives the beak enormous leverage and it is not surprising that they can crack the hardest nuts, most upper beaks have ridges on their curved underside that increase the beak's grasping ability. The beak is also often employed in climbing, serving the parrot as an extra toe. The first and fifth toes of the parrot's foot are permanently directed toward their back, which allows them to firmly grip tree branches.

By and large parrots nest in cavities, with only a few building conventional nests. They lay one to ten eggs (depending on the species) which the female incubates. The young are blind and nearly naked when they hatch. The parents feed the fledglings pre-chewed food from their crop for several weeks until they can leave the nest. And very much like the other two bird species discussed in this book, the condor and the whooping crane, parrots almost always form permanent pair bonds. Such bonds occur even among like-sex specimens when proper mates are not available. Thus it is not in the best interest of a bird lover to keep single parrots as they are social animals and can become extremely lonely. They also reach a good age, some having reached documented ages well over 100 years.

"Amazones," as Puerto Rican parrots are also known, are stocky birds with blunt tails. They are the most famous of the "talking" species. There are 26 species of these parrots and they are distributed over the lesser Antilles. They represent a very old group of birds that fly rather clumsily and instead waddle on the ground. They are perhaps best adapted to climbing trees. They are all green (with minor variations of their head feathers).

Amazona vittata ("vittata" indicating its "banded" head feathers) was down to only 13 birds in 1975 and its extinction seemed imminent. Because of extensive conservation efforts there are now 24 in the wild and 29 in captivity. A degree of optimism now exists but the obstacles to its survival remain formidable.

At one time the bird was abundant in Puerto Rico and its smaller adjacent islands. A principal cause for its decline was, as is often the case with endangered species, the rapid expansion of the human population and the consequent deforestation. Indeed, the vittata's preferred nesting tree was considered to be a weed and was deliberately cut down until quite recently. By 1940 the bird existed only in the Luquillo forest, which has since become a national park. Other perils to these birds are competition for nesting sites with the pearly-eyed thrasher, *Tuxostoma rufum*, interference by swarming honey bees, and a troublesome infection by the larvae of the warble fly.

As with all other endangered species for which a rescue program has been successfully mounted, progress came only when the biology of the species became known through intensive research. Their preferred nesting tree was identified and saved. The ideal depth of nesting holes was studied—that of the pearly-eyed thrasher is shallower—the cavities were better protected and the nest sites prepared and outfitted. Pest strips were affixed to the roofs of the nests to limit the warble fly infection, and other predators, such as rats, were combatted.

An early problem though, in our ability to care for the species, was our inability to differentiate the sexes of these "monomorphic" (meaning the male and female *look* the same) birds. A "pair" donated to the breeding program by the Puerto Rico Zoo eventually was discovered to be a pair of bonded females. Such mistakes could have tragic results when so few birds were left. Thus an intensive program was launched to learn how to differentiate the sex of the birds. In other monomorphic birds, such as the condor and the crane, chromosomes identified from blood cells grown in tissue culture could be used to make the determination. But this method turned out to be ineffective for parrots. Laparoscopy allowed the identification of ovary and testis but the stress of the procedure killed two birds and so was also ruled out as an appropriate method.

Enter the "fecal pellet method," the invention of Bill Lasley and Nancy Czekala of the San Diego Zoo. They argued that one can reliably determine human sex by the steroid hormones found in blood and urine, so why not in birds? The urine-like body wastes of birds come out with their intestinal waste, in the fecal pellets. So they began extracting feces of roosters and hens whose sex was already known. Then they tried it on the waste of cockatiels whose sex was obvious by their plumage. Estrogen and testosterone were measurable in adult birds but their values overlapped. But Bill Lasley conceived the idea of expressing it as an estrogen/testosterone ratio and lo and behold the results fell into two classes: high ratios for the females and low ratios for the males. But the method was still clumsy. It required treating samples with our famous "snail juice" to split complexes and weary-eyed technicians had to spend hours counting radioactive samples. But it worked—at least in adults. After all only adults secrete enough sex hormones to be so scrutinized and even at that they are seasonal in secreting them. Only during certain times of the year will their gonads produce enough steroids to be measured—and you must therefore know the season, and the bird must be healthy. Perhaps he even needs to be "turned on." So it is not easy! But it was done, and Bill Lasley received the prestigious Rolex Award for Enterprise for his discovery.

Arden Bercovitz then took on the project, streamlined the sexing method, and has recently looked into the possibility of sexing the remnants of incubation that stay in the shell after hatching. After all, the sex organs are shaped by sex hormones in embryonic life and excrements accumulate in the egg. And it turned out that in condors this was feasible.

The Parrot Recovery program began in 1968 with help from the World Wildlife Fund. The first parrot biologist was then employed by the U.S. Fish and Wildlife Service. A captive station was begun at Patuxent, Maryland but its birds have now merged with those of a larger aviary built in the Luquillo forest. Here Hispaniolan parrots and the captive Puerto Rican parrots are kept, intensively studied, banded, artificially hatched, and cross-fostered. A population is being built up for gradual release. Attempts at artificial insemination are underway, better nesting methods were developed, and every effort is being made at the same time to study the animals in the wild, which is a very difficult task. It is now known that three eggs are normally laid, that incubation lasts 25–27 days, that the young leave the nest at 60–70 days but remain in the family unit for some time, and we have a more accurate knowledge of what to feed them. However, hawks continue to be a real danger as the area has one of the heaviest red-tailed hawk populations on the islands. But with the aid of such techniques as double-clutching and cross-fostering eggs with the Hispaniolan parrots, both of which have been successful, the first real ray of hope appeared when, in 1979, the program successfully hatched and reared its first chick in captivity.

Further Reading

Benirschke, K., ed. Primates: The Road to Self-Sustaining Populations. Springer-Verlag, New York, 1986.

A comprehensive review of the state of apes and monkeys in the world, their scientific contribution to man's health problems, and considerations on how to save them for the future.

Cadieux, C. These Are the Endangered. Stone Wall Press, Washington, D.C., 1981.

This is a sympathetic presentation of North America's most threatened animals, the efforts to save them, the agencies concerned with their survival, and the role national parks play in the conservation.

Ehrlich, P. and Ehrlich, A. Extinction. Random House, New York, 1981

This is the most comprehensive treatment of the process of extinction, the politics, causes, and conservation issues in general.

Gotch, A.F. Mammals—Their Latin Names Explained. Blandford Press, Poole, Dorset, 1979.

This book gives fascinating insight into the discovery of mammals and how they acquired their names. A similar book on birds exists.

Fitter, R. and Scott, P. the Penitent Butchers. Fauna Preservation Society, London, 1978.

This small volume gives the venerable history, goals, and achievements of the oldest (1903) conservation society in readable detail.

Hearn, J.P. and Hodges, J.K., eds. Advances in Animal Conservation. Oxford University Press, Oxford, 1985.

Prince Philip, President of the World Wildlife Fund, introduces this recent overview of zoos' contribution to our wildlife conservation efforts.

Martin, E. and Martin, C.B. Run Rhino Run. Chatto and Windus, London, 1982

This remarkable book concerns not only the rhinoceros' decline but exemplifies the problems faced by all large animals.

Michael, R. The Audubon Society Handbook for Butterfly Watchers. Scribners & Sons, New York, 1984.

Thibodeau, F.R. and Field, H.H. eds. Sustaining Tomorrow. University Press of New England, Hanover, New Hampshire, 1984.

This small volume is must-reading for anyone with an interest in the strategy for world conservation and development.